CHEERS!
WINE
CELLAR
DESIGN
III

ARTPOWER

Cheers! Wine Cellar Design III
© Artpower International Publishing Co., Ltd.

ACS ARTPOWER

Edited and produced by Artpower International Publishing Co., Ltd.
Address: Flat B5 1/F, Manning Ind. Bldg., 116-118 How Ming St., Kwun Tong, Kowloon,
Hong Kong, China
Tel: 852-23977886
Fax: 852-23982111

www.artpower.com.cn
contact@artpower.com.cn (Editorial)
book@artpower.com.cn (Sales - China)
overseasales@artpower.com.cn (Sales - International)

ISBN 978-1-913536-94-7

Publisher: Lu Jican
Chief Editor: Wang Chen
Executive Editor: Zhou Ziqing
Art Designer: Chen Ting
Cover Design: Xiong Libo

Size: 235mm X 290mm
First Edition: April, 2022

PREFACE

Every wine collector has a dream of building their own wine cellar to house a wide range of wines. However, wine cellar design is not an easy task, as it requires a number of factors to be taken into account and a comprehensive understanding of wine collecting culture. For example, the temperature, light, humidity and vibration disturbances in a wine cellar make the design of a wine cellar a little more special than other projects. A good wine cellar design must therefore ensure that the wine is firstly of good quality, followed by a variety of clever designs incorporated into it.

This book is the third in a series of books on wine cellar design, featuring a selection of recent wine cellar examples from a number of outstanding design agencies from home and abroad. Each project is accompanied by a professional interpretation of the text, as well as detailed drawings and floor plans. The book contains a number of tips and notes on wine cellar design, from a holistic point of view as well as sharing experiences in the smallest details.

What is exciting is that many of the wine cellar design projects included in the book are innovative, unconventional and eye-catching. These examples represent the latest global trends and can be of benefit to both the design agencies involved and the collectors who dream of having their own wine cellars. We expect these wine cellar projects to inspire designers to create more stunning designs that will break the stereotype of what a wine cellar should be. Now that wine cellars are part of the trend, how can we leave them in their monotonous and uninteresting state?

The publication of this book is thanks to each of the design agencies that have contributed to our projects, as you have made it more informative and attractive. We also look forward to the publication of further books in the Wine Cellar Design series, which means that wine cellar design is taking more centre stage and we will see more fabulous designs!

Artpower Editorial Office

CONTENTS

De Vinos y Viandas

The concept of this space comes from the searching of a well-known world which is the wine and its sales spaces. There are a lot of these references, thus, designers attempt to abstract these images in a manner which the circumference shape arises.

The circumference shape is very noticeable in many ways, for example, in the wooden barrels that can be visible inside the cellars, in the bottles, in the antique vaulted cellars. Therefore, the circumference introduces customers into this world as the leitmotiv of the project.

The juxtaposition of several circumferences in transversal and longitudinal direction of the local creates a space where the required situations appear such as the counter, tasting table and exhibition-sale space.

This chain of curves originates a series of arches of the space that linearly will be seen as vaults, a constructive system that evokes the ancient underground cellars.

Once the system is generated, designers atomize it in its longitudinal direction dividing the system in vertical spaces of 15 cm, obtaining in this way two frontal elevations of product exhibition. Therefore, the space is perceived as a sequence of wooden ribs in which the circumference shape is even more present.

Design Agency: Zooco Estudio / Team: Javier Guzmán Benito, Miguel Crespo Picot, Sixto Martín Martínez, Beatriz Cavia Peñalva, Beatriz Villahoz Herrero, Elvira Segovia Moreno, Jorge Alonso Albendea, Juan García-Segovia, María Larriba Hombrados, Marta Borrás Quirós, Miguel Montiel Martín, Paula Cruz Fernández / Location: Valladolid / Area: 80 m² / Photography: Imagen Subliminal

Optimum viewpoint to enjoy the impressive circular elements.

Floor plan

Counter

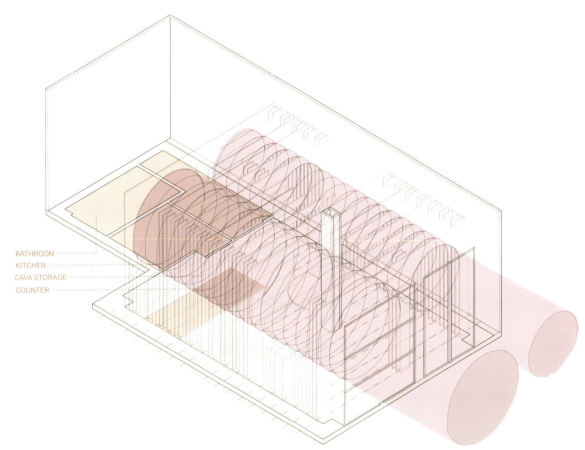

BATHROOM
KITCHEN
CAVA STORAGE
COUNTER

Chain of curves

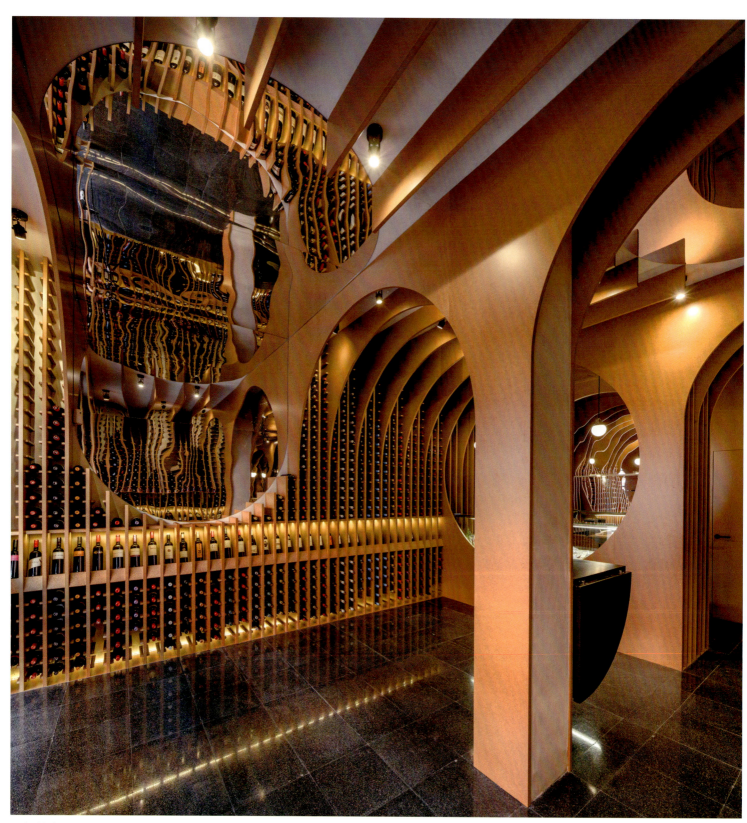

Embedded mirror enlarges the space.

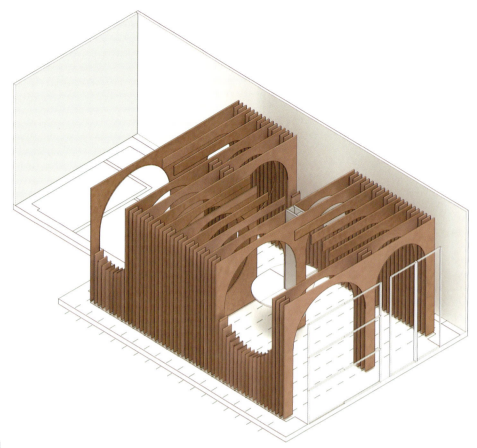

Elevation 1

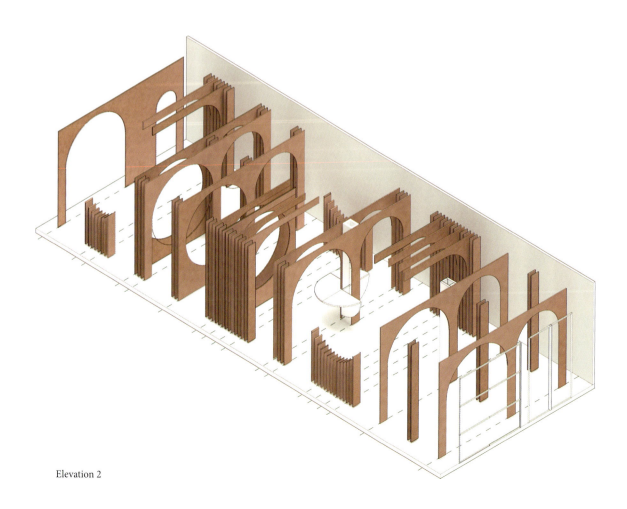

Elevation 2

Making the racks become irreplaceable decorations.

Step into the world of circumference.

Sky-Cellar Beijing

Superimpose designed an exclusive members club and hidden wine cellar for a wine enthusiast. The project is situated at the top floor of one of Beijing's newest and most high-end tower developments, the Genesis Community. Beijing Genesis is a mixed-use development, designed by Japanese architect Tadao Ando, combining a hotel, offices, gardens and a museum. The members' club will be open to private individuals and selected groups of members only.

The members club is where clients organise gatherings, store and consume their valuable wine and tea collections. The space is divided by a central element creating two distinctive worlds: the client's office space and the private member's club. Members enter through a hidden pivot-door into the exclusive member's club that instantly offers spectacular views over Beijing's embassy and financial districts. The golden stainless steel central element subtly reflects the skyline and naturally forms a backdrop of the entire members club.

A segment of the central divider is constructed with dark mirrored glass which mirrors the skyline and hides the "Sky Cellar". The design of this "Sky Cellar" is a reinterpretation of the traditional wine cellar. Members access into the hidden exclusive wine cellar through the automated sliding glass doors. The 25 m^2 cellar stores up to 500 bottles of wine from six famous wine regions around the world: Bordeaux, Rhone Valley, Burgundy, Napa, Mosel and Barolo. During day time, the mirrored glass doors hide the wine cellar and protect the wines from direct sunlight. The mirror doors allow members to admire both the wine and skyline whilst being inside the wine cellar. At sunset and during evening hours, the wine cellar lights up and reveals the 500 wine bottles to the main space.

Design Agency: Superimpose Architecture / Design Team: Carolyn Leung, Ben de Lange, Ruben Bergambagt, Huimin Xie, Yujia Deng, Xiaoyu Xu, Casper Kraai / Location: Beijing, China / Area: 203 m^2 / Photography: Marc Goodwin and Superimpose
Materials of Wine Rack: Acrylic, Stainless steel

The hidden pivot-door, view from the exclusive member's club to the entrance lobby.

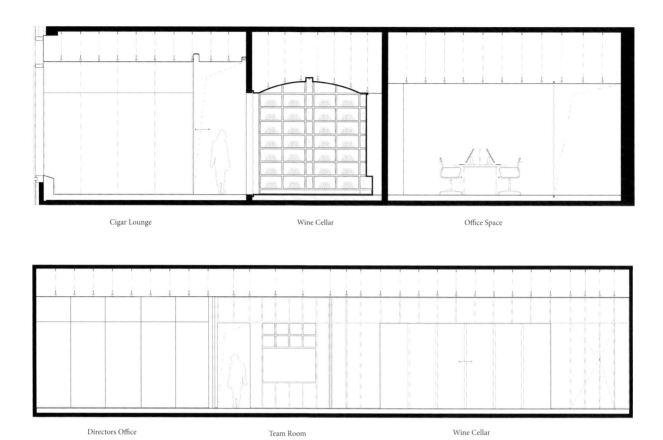

Cigar Lounge

Wine Cellar

Office Space

Directors Office

Team Room

Wine Cellar

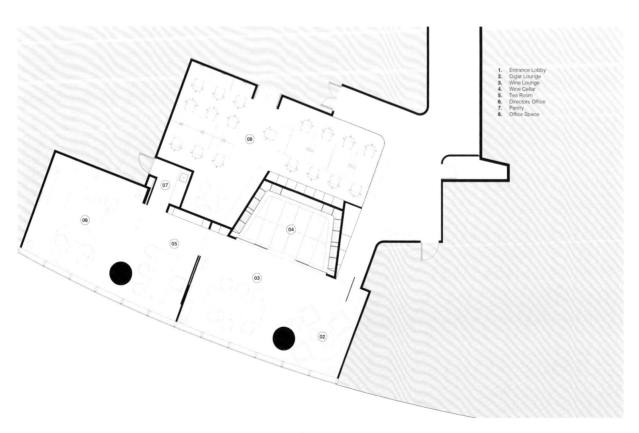

1. Entrance Lobby
2. Cigar Lounge
3. Wine Lounge
4. Wine Cellar
5. Tea Room
6. Directors Office
7. Pantry
8. Office Space

Floor plan

The mirrored glass doors hide the wine cellar and protects the wines from direct sunlight.

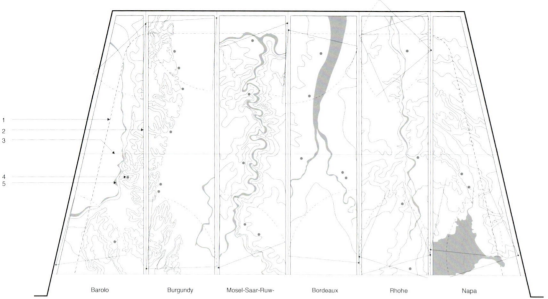

Wine map

An abstract representation of the six wine regions.

The stainless steel shelves, an acrylic base.

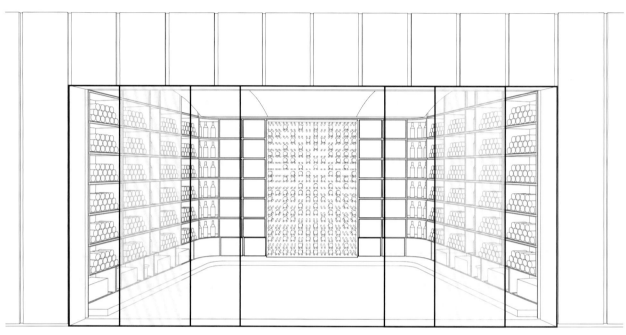

Section - Wine Cellar 1

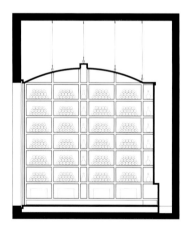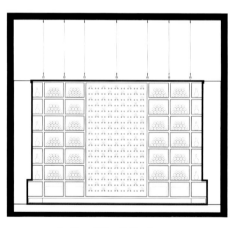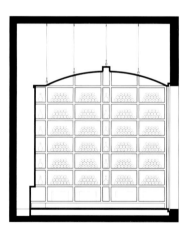

Section - Wine Cellar 2

An abstract representation of the six wine regions is installed underneath a glass floor with natural light. Superimpose designed the maps and constructed them by using concrete ash mud for the topography and stainless steel for the rivers. Within the stainless steel shelves, an acrylic base is being used and designed in such a way that it allows for multiple storing positions of the wine bottles. Bottles can be displayed and stored either stacked or inclined to display the bottle etiquette. At the same time the acrylic base allow for an optimised light distribution inside the wine cellar.

Golden steel racks give a sense of luxury.

Wijn aan de Kade

Rotterdam based Studio Aaan has had the fortune to combine architecture and wine in their design for the new wine shop "Wijn aan de Kade". It is located in an 1896 monumental premises along the Admiraliteitskade in the centre of Rotterdam. The four storey shop has been transformed into a long, modern and warmly finished space clad with laser engraved oak panels.

The continuous space is flanked by two cabinet walls with niches, cupboards and amenities. Wine boxes are stored inside the cabinets, large cut-outs function as displays for over 300 different bottles of wine. Service areas like storage, toilets and an office are concealed behind the wall. The result is a clear and spacious shop with a strong focus on the wines. The cabinets are clad with oak panels which are engraved with labels of famous and desirable wine houses. The rear wall of the wine display cut-outs shows the original brickwork of the foundation of the seventeenth-century VOC shipyard (Dutch East India Company).

The 200 m^2 shop, consists of four level: a basement, ground floor, mezzanine and the first floor. In addition to the main retail space, the shop features a small bar, a kitchen, a tasting area and a cellar. The lowly lit wine cellar has an intriguing atmosphere while the other floors are bright and generous. Careful integration of daylight from above balances the perception of the entrance and the mezzanine in the back.

All furniture is custom designed. In the cellar two long black cabinets filled with fine gravel contain luxurious wines. A large cabinet with integrated cash register is placed on the ground floor. A small bar is placed on the mezzanine to accommodate small tastings. The first floor is used as the main tasting area for larger groups and contains educative tables with maps of wine regions engraved in the table tops.

Design Agency: Studio Aaan / Client: Bruderer Property b.v. / Photography: Sebastian van Damme, Adriaan van der Ploeg

There are over 300 different bottles.

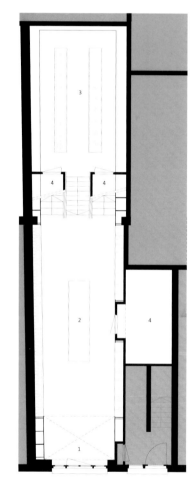

Basement / Ground Floor

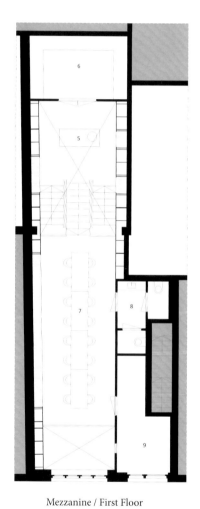

Mezzanine / First Floor

1   entrance
2   wine shop
3   wine cellar
4   storage room
5   bar
6   kitchen
7   wine tasting room
8   toilets
9   office

0   2

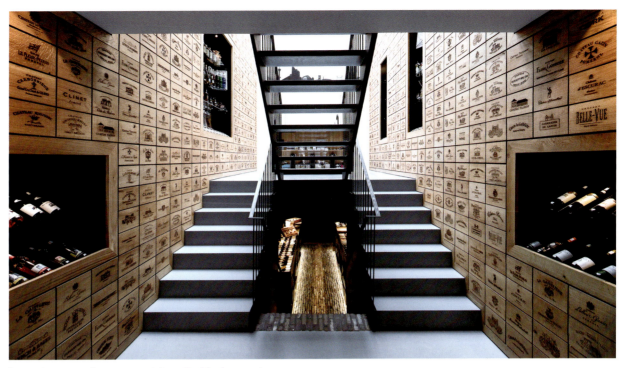

Impressive contrast between grey stairs and golden basement.

Wine shop and bar, seen from the mezzanine.

Tasting area on the first floor with several educative tables.

Double pendant lamps above the entrance.

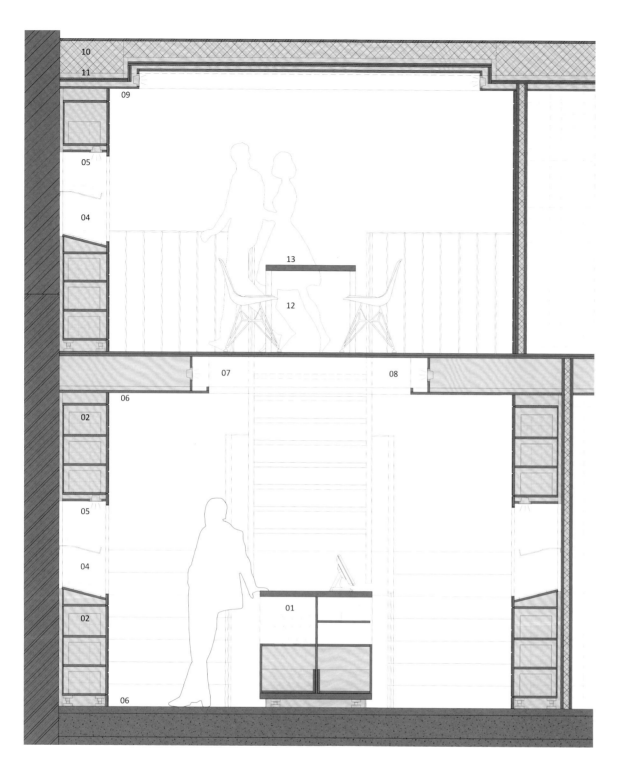

01 cabinet plywood 18mm (black) oak drawers 18mm
    cashregister, storage and presentation
02 cabinet MDF 18mm (black), doors decorated with oak panels (6mm)
    storage for 6 wine boxes
03 cabinet MDF 18mm (black), doors decorated with oak panels (6mm)
    storage 20 wine boxes
04 wine display
05 brick wall spray painted (black)
06 oak moulding (6 mm)
07 spotlight, LED 3000K (black)

08 beams pinewood 100x280mm (black)
09 decorative ceiling plywood (white)
10 soundinsulation 140-280mm
11 fire resistand ceiling (60 minutes)
12 chair Eames DSR, (black)
13 table, steel frame oak tabletop with engraved maps of
    different wineregions

Section 1

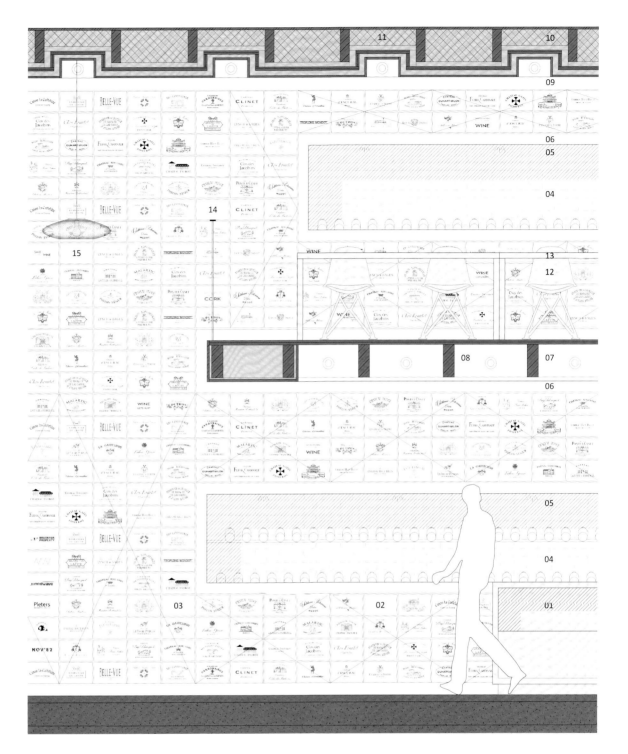

Section 2

01 cabinet plywood 18mm (black) oak drawers 18mm
    cashregister, storage and presentation
02 cabinet MDF 18mm (black), doors decorated with oak panels (6mm)
    storage for 6 wine boxes
03 cabinet MDF 18mm (black), doors decorated with oak panels (6mm)
    storage 20 wine boxes
04 wine display
05 brick wall spray painted (black)
06 oak moulding (6 mm)
07 spotlight, LED 3000K (black)

08 beams pinewood 100x280mm (black)
09 decorative ceiling plywood (white)
10 soundinsulation 140-280mm
11 fire resistand ceiling (60 minutes)
12 chair Eames DSR, (black)
13 table, steel frame oak tabletop with engraved maps of
    different wineregions
14 steel railing
15 fiber pattern lamp, Ufo series, Atelier Robotiq

In the low-lit basement, luxurious wines are stored with fine gravel.

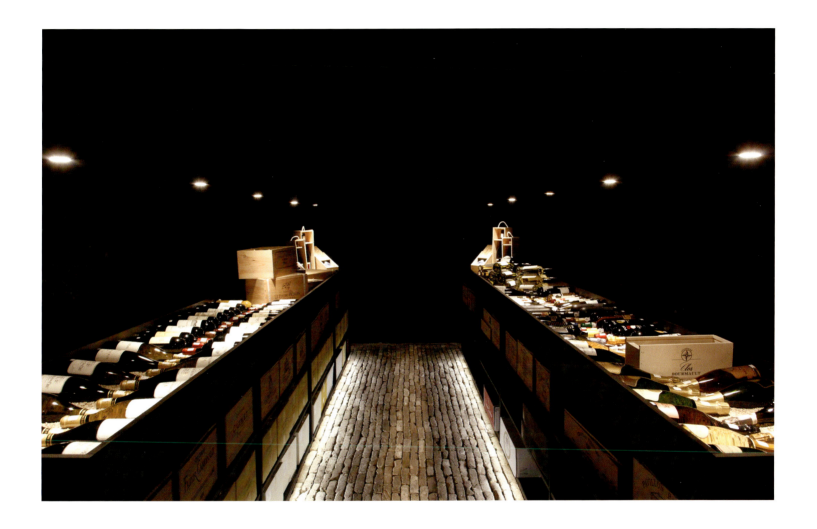

Hotel Hubertus

The Hotel Hubertus is located in Valdaora, at the foot of the famous ski and hiking area Kronplatz in the Puster Valley at an altitude of about 1350 m. The family establishment was generously enhanced and enlarged with 16 new suites, a new kitchen with restaurants and "Stuben", an entrance area with lobby, reception and wine cellar and fitness and a relaxation room with panoramic terraces. The new 25 m long pool, functioning as a connector between old and new, underlines the essence of this comprehensive renovation and renewal project.

The key challenge in the project was to create a link between the existing building and the new design, in order to keep a uniform and consistent appearance. Through the use of a warm, grey brown earth tone, derived from the colour palette of the surrounding landscape, a uniform design was achieved.

Architects: noa* / Interior Design: noa* / Location: Valdaora, Italy

View from the entrance.

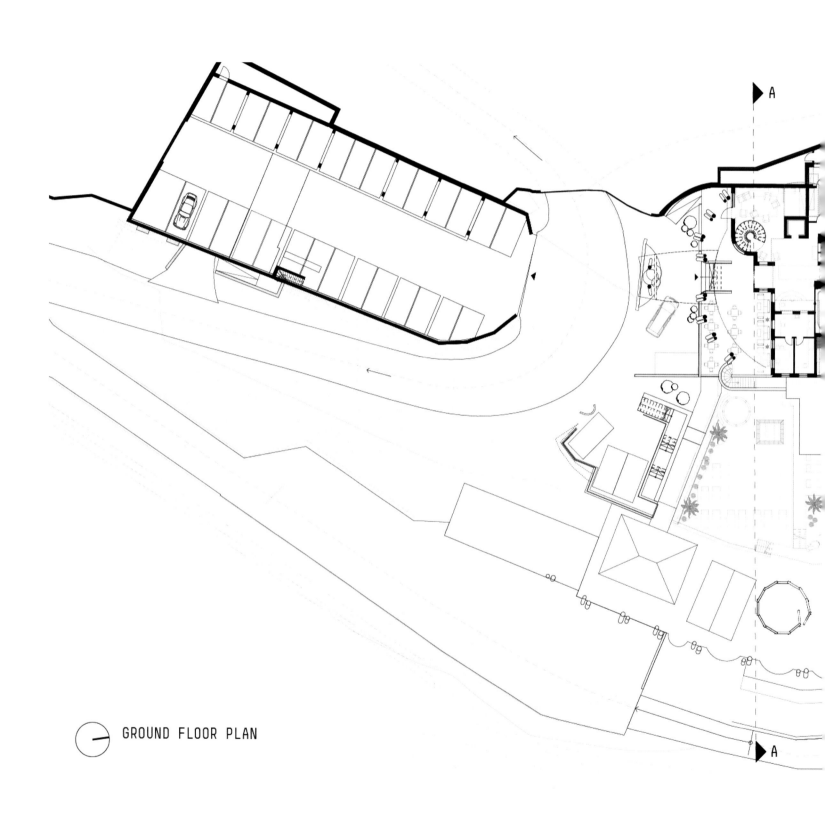

GROUND FLOOR PLAN

A

A

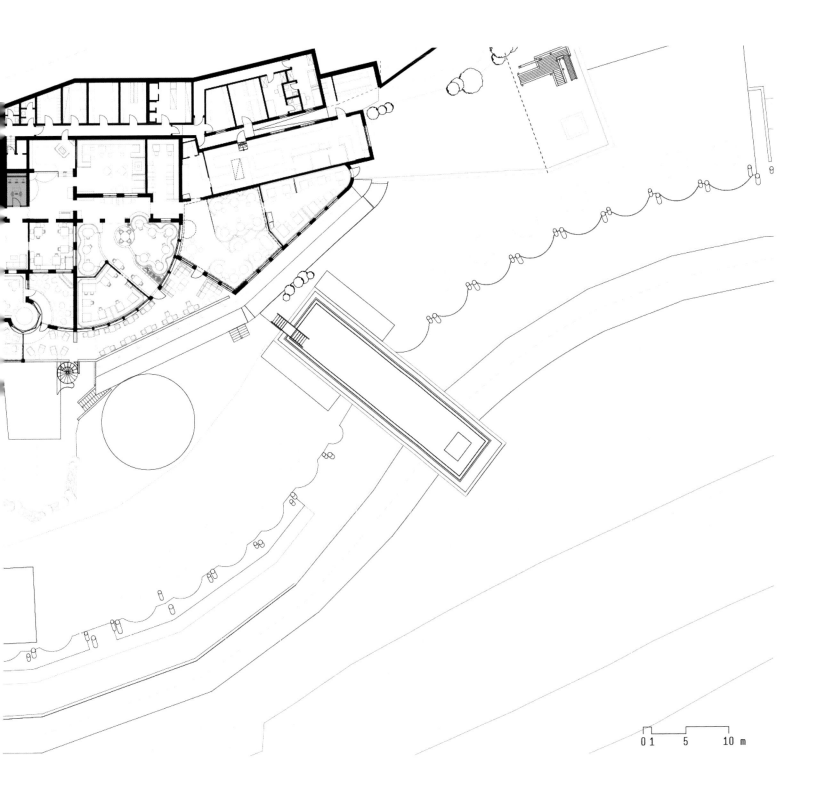

0 1   5   10 m

Mirrors are installed in both sides to enlarge the room.

Only one platform in this room.

Zallinger

This wine cellar is designed for Zallinger Hut which is over 160 years old. The design agency noa*, an architecture studio in South Tyrol, has long been committed to developing innovative models of receptivity and green tourism. The aim was to improve the quality and accommodation capacity of an old high mountain hotel structure without altering the delicate landscape and environmental balance, while at the same time creating aesthetic value and sustainability.

Sustainability, respect for the mountains and direct contact with nature are the principles that have guided all design choices. For example, in the design of the new rooms, careful use of the space was made to provide high levels of comfort in relatively small sizes; the roofs wood shingles, typical of the South Tyrolean tradition, were used; all the materials have been certified and the complex of buildings, heated by pallets, has obtained the Clima Hotel certification.

Design Agency: noa* / Location: Italy / Photography: Alex Fil

Elongated compartments accommodate the individual bottles of wine.

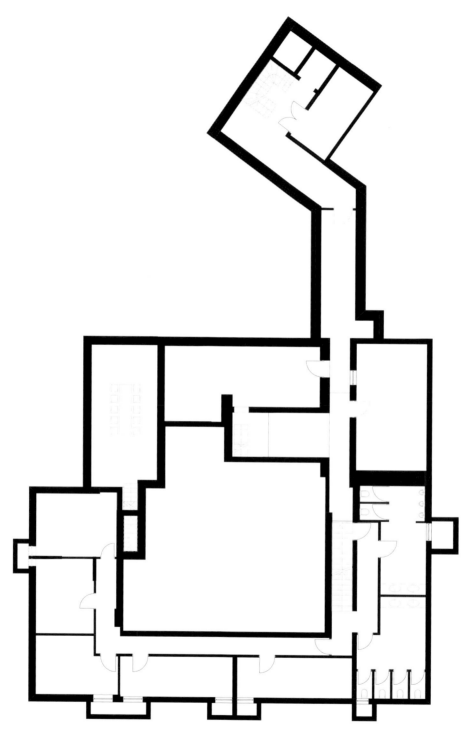

Wine cellar

View of the wine cellar to the entrance.

Front view for the "bookcase".

The wooden shelves create a sort of bookcase,
characterized by inclined upright crafts of wood.

Vigneti Le Monde

Le Monde winery nestles between the sea and the hills of North East Italy, an area steeped in traditional wine culture. The focus of this wine tasting room was to create a space devoted to sensory experience that ensured a constant visual connection with the surrounding vineyards.

To enhance this connection, two large tasting tables extend in line with the rows of vines that stretch as far as the eye can see outside, merging the long linear landscape of the vineyards with the wine tasting room's interior. The pair of tables are 9m, solid, textured walnut, divided in the middle to reveal an inner contrasting Corten core. The cut in the center has a variety of functions, including display of bottles, food or a place for spittoon containers during a tasting session. The core also has structural purposes, holding the two sides of the table together. The tables, despite their size have only two points of structural support where they meet the ground, one of which folds up to create a standing area. The engineering of the table was challenging as it required floor reinforcement and an additional metal floor plate on which the table would be fixed once delivered on site.

A bespoke bottle display from Corten and timber and a main counter that works as both kitchen and feature bottle display are placed at the front of the room as one enters. The 7m in length and 2.7m in height bottle display was designed to offer maximum flexibility, holding; Two sub-zero wine fridges, and adjustable boxes of different sizes of Corten and timber used as storage or display area.

The 4-meter-long counter functions as a kitchen as well as a display area, which surface is tilted, as if a corner of the block has been removed to present a corten, laser-cut, illuminated map of the vineyard and a bottle display.

The overall result is a room that is at one with its surroundings and offers visitors a beautifully considered environment from which they can sample the produce of this fine Italian winery.

Interior Design: Alessandro Isola / Design Team: Alessandro Isola Studio / Location: Italy / Client: Vigneti Le Monde
Photography: Studio Auber / Materials of Wine Rack: Wood, Corten, Stone

Bespoke bottle display with flexible modules from Corten and timber, 7 m long and 2.7 m high.

The light installation produced by Vibia, is reminiscent of pouring wine, creating a connection between the staircase and the counter, gently encouraging visitors through and up the main space.

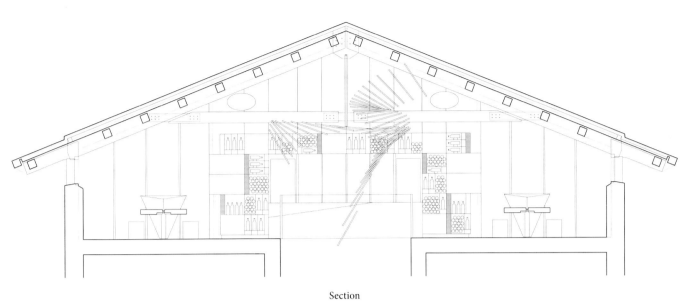

Section

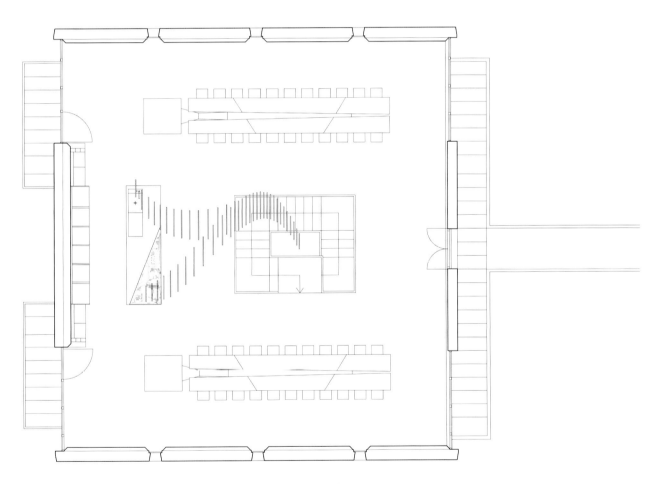

Floor plan

The main counter, made of stone, is engraved with the winery's logo and map.

The stone counter is used as bottle display as well as a kitchen.

Wine Cave

Winner, 2018 NSW AIA Commendation in Interior Projects Award, Wine Cave.

The site lies under an Edwardian mansion, which also has a pre-war bomb shelter under its front lawn. The client who had been working directly with a builder and engineer wanted to take the opportunity to renovate and repurpose the space under the house. It quickly became apparent that the original block-work was at an incline, sloping with the site, and that the existing topography outside of the room, could still be read inside the room. There was a poetic feel to the persistence of this permanence when all else was in a state of flux.

MWA proposed a new concrete underpinning setback from the edge of existing stone blocks enabling vertical sandstone panels to be laid out under the length of the original foundations. The inclined horizontal of the now suspended stone foundation blocks could be clearly revealed and understood. The result achieved was the creation of 3 formal rooms from the original foundation stone work. An original cellar, new dining room and gallery and addition of a new bathroom. A new entry hatch and stair was added, providing a formalized and easier connection from the main house to cellar.

Interior Design: McGregor Westlake Architecture / Design Team: MWA - Wes Grunsell & Peter McGregor
Location: Wollstonecraft, NSW / Photography: Katherine Lu

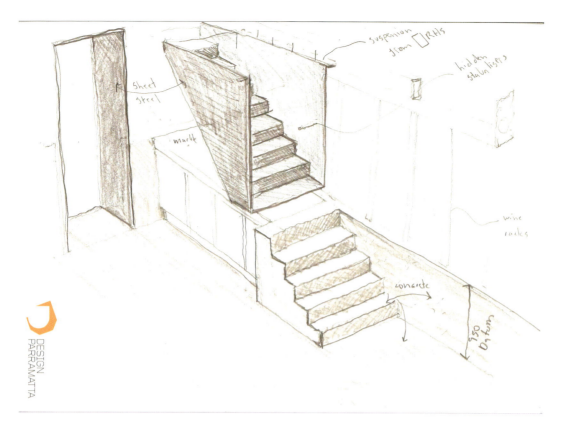

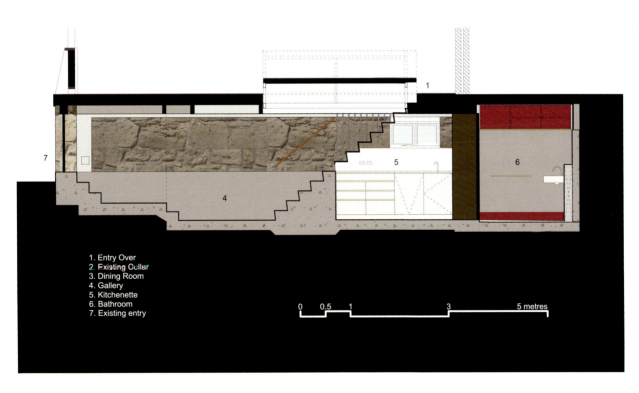

1. Entry Over
2. Existing Cellar
3. Dining Room
4. Gallery
5. Kitchenette
6. Bathroom
7. Existing entry

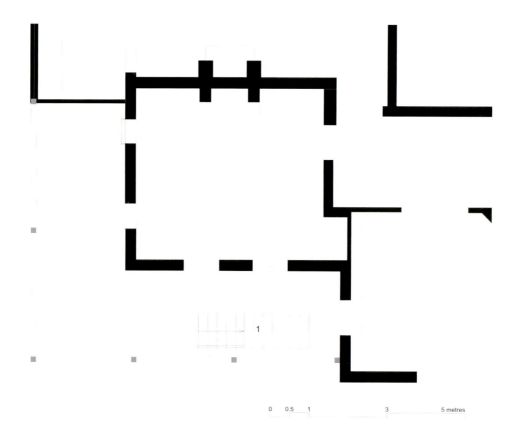

1. Entry Over
2. Existing Cellar
3. Dining Room
4. Gallery
5. Kitchenette
6. Bathroom
7. Existing entry

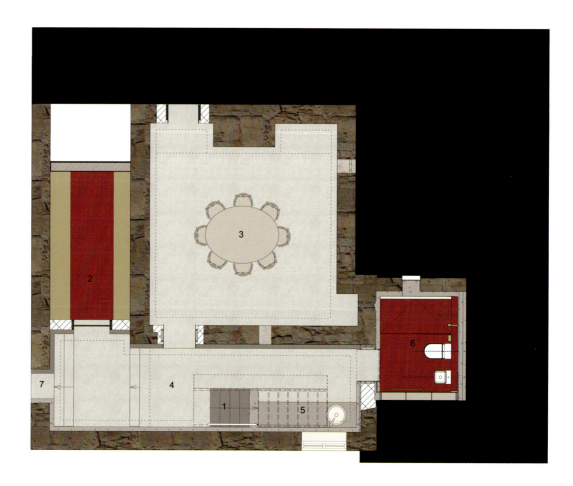

1. Entry Over
2. Existing Cellar
3. Dining Room
4. Gallery
5. Kitchenette
6. Bathroom
7. Existing entry

Delas Frères Winery

The rocky, terraced, undulating hills above Tain l'Hermitage facing south are renowned as some of the best terroir along the Rhone Valley. Delas Frères have cultivated some of the most select parcels since the last century and were determined to create a modern winemaking installation while investing in the centre of the historic town at the foot of the hills. It was a decision to reaffirm the identity of the winery and the "côteaux", or hills, despite the challenges of wine harvesting in an urban environment.

The heart of the new winery is a historically walled garden. Using solid stone construction, the new wine chai and shop become walls framing a renovated manor house and its park, tying the site to its context. Visitors wander through the garden past the wine shop on the street, and through the chai to the manor house.

The chai forms an undulating garden wall in stone, built for quality and emotion. The stone relates to the site, while the thermally inert, porous walls create ideal conditions for wine. Ramps within the chai allow visitors to discover the wine process within an efficient concrete interior, and lead to views to the hills from a roof terrace, and down to the bottle cellar under the manor house.

The shop forms the opposing garden wall, a linear space behind shading, staggered stone pillars. An existing chestnut tree traces a bite out of the wall, under which one finds the shaded, glazed entrance of the shop. The existing mansion affirms itself as the central element of the garden, and is renovated as a guest house, linked to the chai. It has a restaurant and tasting rooms, bedrooms overlooking the garden and a cellar for the historic bottle collection.

Architects: Carl Fredrik Svenstedt Architect, with Carl Fredrik Svenstedt, Boris Lefevre, Pauline Seguin, Thomas Dauphant, Marion Autuori, Benoit- Joseph Grange / Interior Design: Carl Fredrik Svenstedt Architect, decorator / Location: France
Area: Chai 3,200 m², Wine shop 400 m², Guest house 1 400 m² / Client: Champagne Deutz Delas Frères
Photography: Dan Glasser / Materials of Wine Rack: Wood (Custom built wooden shelves and storage)

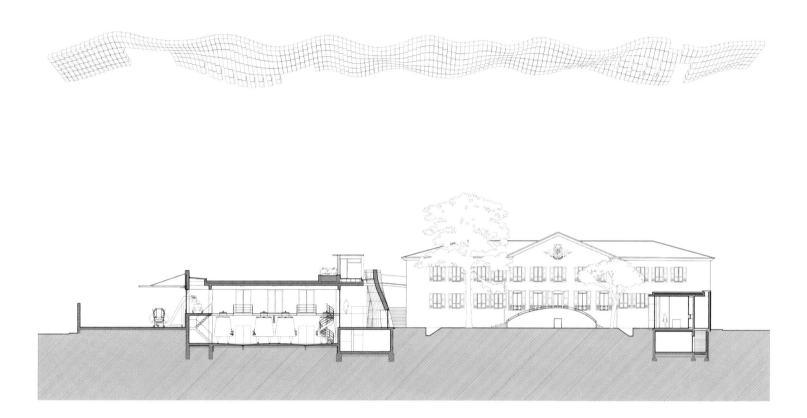

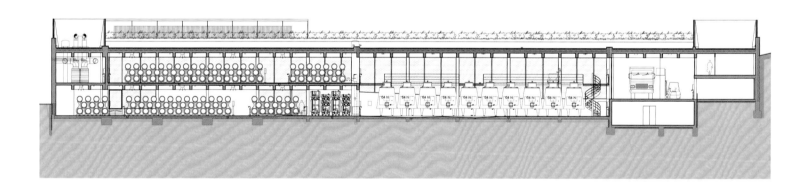

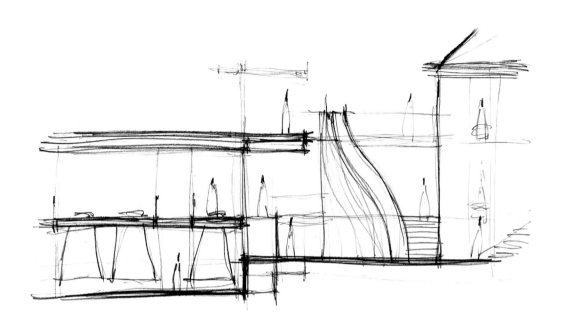

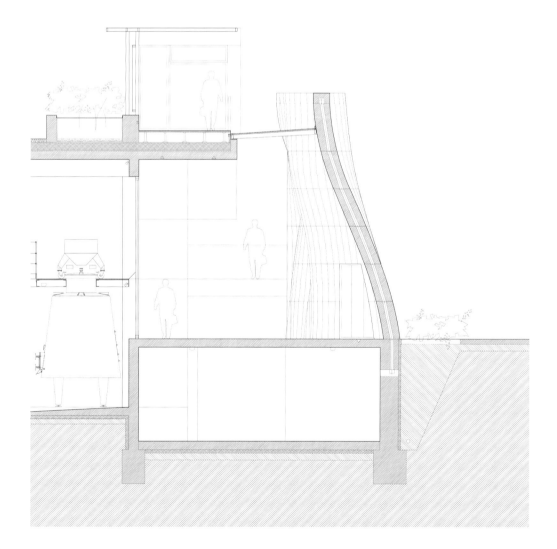

This chai is built to be touched. The structural facades are made of load bearing, fifty-centimetre-thick Estaillade stone from down the Rhône river. The tender, relatively light sandstone is ideally adapted to massive stone construction, being workable and best in thick structural blocks.

The main, undulating wall is made from blocks carved by robot, which are post-tensioned to the foundations and bonded horizontally using stainless steel cables. The blocks are individually cut by wire from pre-dimensioned rectangular blocks and then shaped by a robotic router. Intelligent machining reduces waste, while the resulting gravel is reused to pave the garden. Despite the unique technicity of the wall, the blocks are mounted traditionally by a two-man father and son team of stonemasons using hand tools and a crane.

The foundations, floors and large spans are made of concrete. The undulating wall is eighty metres long and seven metres high, with a geometrically stable, structural form.

Sunlight enters the visitors' space behind the wall through a continuous skylight, the wall serving as a light reflector for the tank and barrel halls, where direct light would be detrimental. The undulations of the wall define the spatial sequence for visitors in this "déambulatoire", creating places of arrival and for viewing the wine process. The ramped space ties together all the existing and new levels of the site, from the garden to the rooftop belvedere and the views to the vineyards.

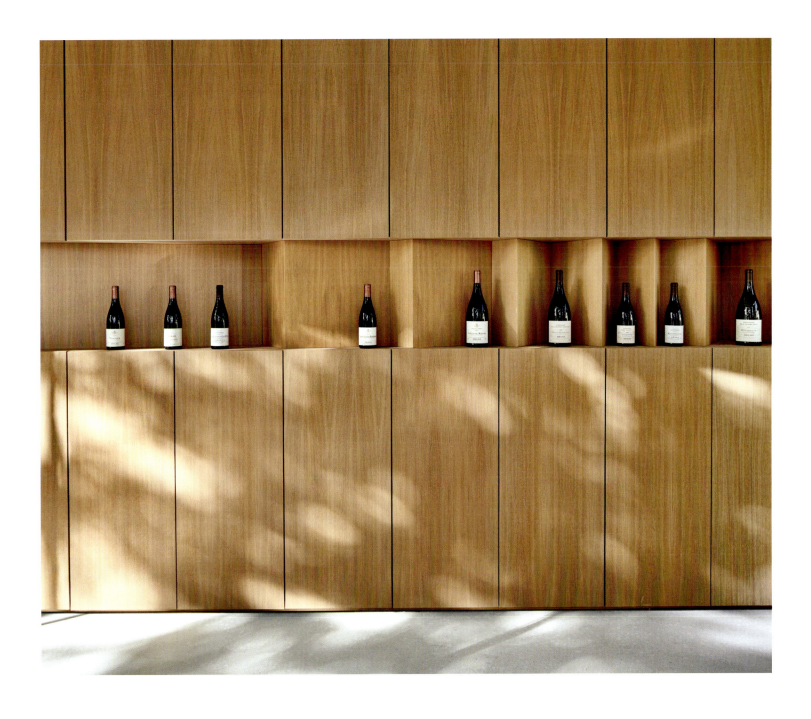

Domaines Ott Winery

Building in stone implies carving a mountain, the result imposing and profound, creating a presence with self-evident materiality. On this site, near the Cistercian Abbaye du Thoronet, a building with stone extracted from Roman quarries places the project in a temporality resonant with the landscape.

The stone blocks, mathematical, are one by one metres by fifty centimetres thick and weigh exactly one metric ton. They rise in equilibrium ten metres high, twist and turn. The walls dilate, filigrees of pure weight in the sun.

The winery and visitor's centre marks a new horizon in the Provençal landscape, a mineral presence anchored in the rolling vineyards overlooking the historic Chateau de Selle. Two walls in solid stone rise parallel to the road and wine terraces, the one curved to follow the speed of passing vehicles. The massive walls frame the winemaking process, sheltering the wine, work and visitors. The walls are both imposing and light, shifting as needed to become porous screens, providing views, access and ventilation.

The building is partially sunk into the hill, a thermally inert emergence optimised for winemaking. The slope allows for a natural gravitational flow and a coherent linear process, visible from the public esplanade and reception areas overlooking the cask-room and steel tank hall.

The sun warms the surface of the stone, soft as sand. Visitors can measure themselves against the human scale of the blocks, close enough to be touched. It is a meeting of the senses. What remains are the pines, the vines and the mountain.

Architects: Carl Fredrik Svenstedt Architect; Carl Fredrik Svenstedt, Tae In Kim, Camille Jacoulet, Thomas Carpentier, Clement Niau / Interior Design: Carl Fredrik Svenstedt Architect / Design Team: Apart from above mentioned: Christophe Ponceau & Mélanie Drevet (landscape architects) / Area: Main winery building 4,200 m², agricultural building and vehicle shed 850 m² Client: Les Domaines Ott / Photography: Dan Glasser & Herve Abbadie / Materials of Wine Rack: Wood

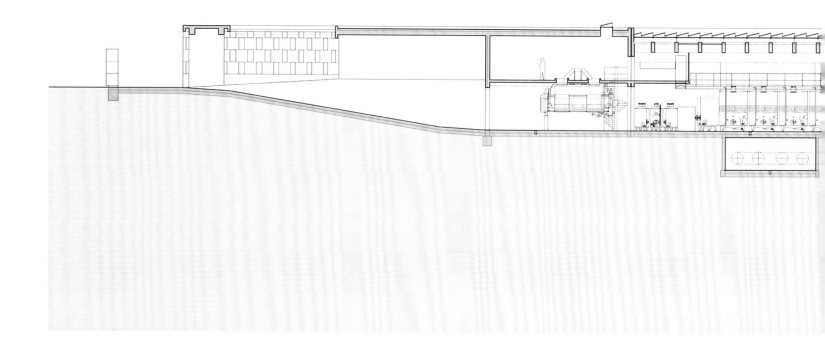

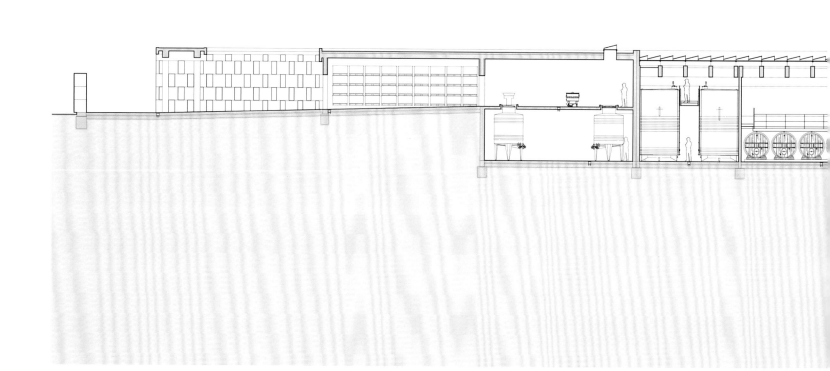

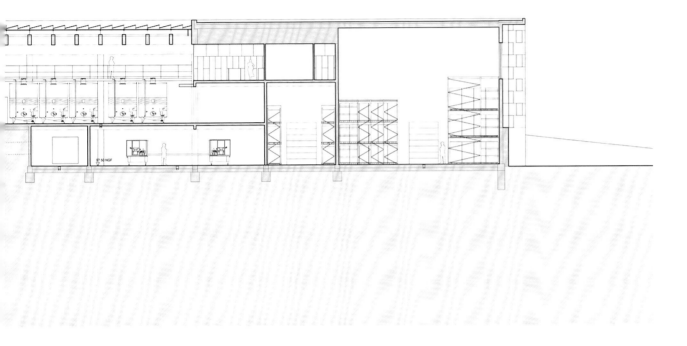

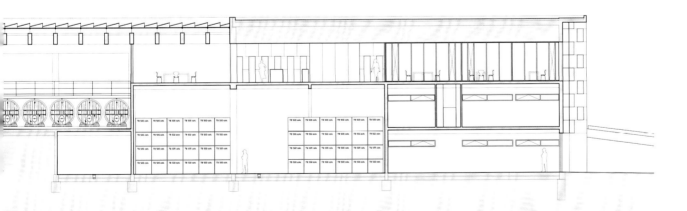

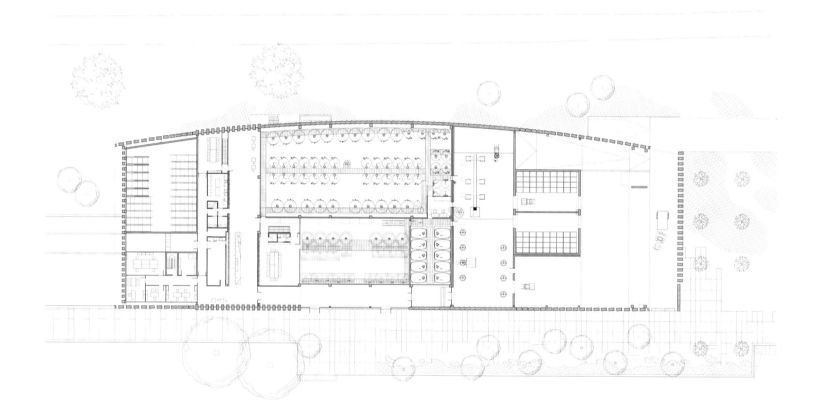

Wine Tourism Building

Located in Penafiel, Portugal, Quinta da Aveleda is one of the main producers of Vinho Verde in the country and enjoys an unusual built heritage surrounded by beautiful gardens.

The reconversion of a former agricultural building (barn) to a wine tourism building is based on the spatial potentialities of the pre-existing building and the creation of a visiting route in three acts. In opposition to the dense compartmentalization originally existing, the building is now organized in a succession of continuous spaces, formally related to the volumetry of the building: the tasting room (complemented by a small kitchen) in one of the side wings; the vinoteca and the mezzanine in the central volume; and the shop in the remaining side wing.

The tasting room emphasizes the longitudinal character of the building through the linear organization of the long tasting tables, which promote up to a hundred seats;

The vinoteca is the space of transition between the tasting room and the shop, also giving access to the mezannine. It intends to be an experience space, dedicated to the exhibition of special wines. Chromatically distinct, this is a space of pause that seeks to create an intimate atmosphere. This space is intrinsically related to the mezannine, which is organized in two elevated spaces and a connecting bridge, allowing a dominant reading of the two adjacent spaces on the ground floor. Lastly, the shop has a central layout of furniture, releasing, like the tasting room, the peripheral walls of the building and thus allowing it to receive, among the spans that regulate its rhythm, photographic and written memories and objects.

Architects: Diogo Aguiar Studio / Design Team: Diogo Aguiar, Daniel Mudrák, Adalgisa Castro Lopes / Location: Penafiel, Portugal / Client: Aveleda, SA / Photography: © Fernando Guerra

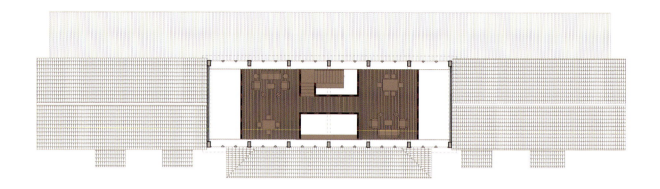

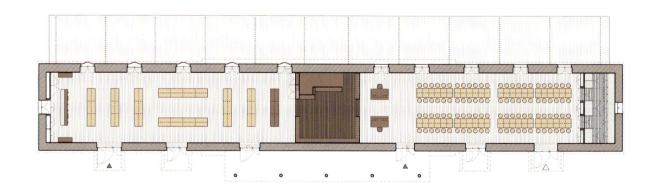

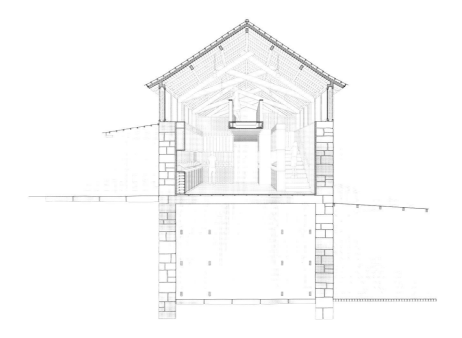

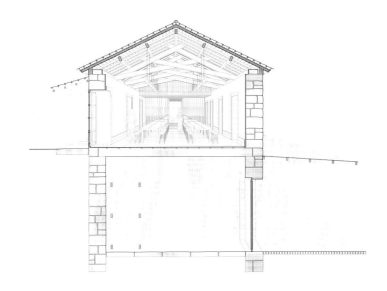

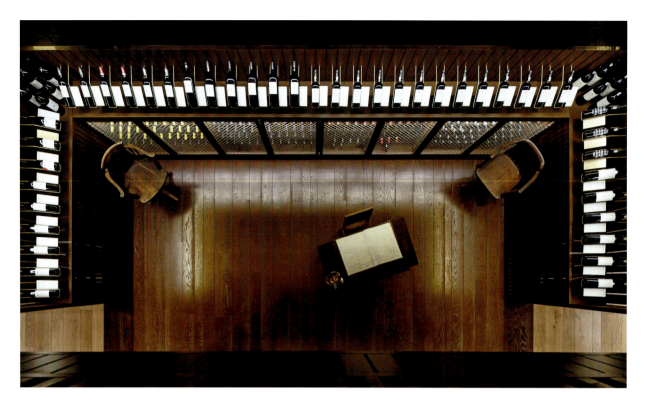

Inherent to the new design of space, is therefore its understanding as museum space as a spatio-temporal opportunity to narrate the history of the family that for five generations produces the Vinho Verde in that region.

Recovering the traditional spirit of painted wood which can be found in many farms in the north of the country, the intervention is based on a classic composition between two colours, the beije and the brown which also existed in other buildings of Quinta da Aveleda, attributing this time a greater drama and contemporaneity by the assumption of a strong chromatic contrast between different spaces in direct dialogue.

At the structural level, due to technical impossibilities, the original trusses of the building were not maintained: those of the central body, the elegant timber scissor trusses, were no longer in the ideal state of conservation; and those of the side wings had already been tampered with armoured brick. In this sense, the less usual design of the timber scissor trusses was recovered initially only existing in the central volume of the building and now also adapted to the lower bodies of the building. At the same time that they liberate a greater height giving the sensation of a greater spatial amplitude, these unusual asses also assume a singular iconography, characterizing this particular building, differentiating it from the others.

Concluding, the whole architectural intervention is based on the respect and the reinforcement of the identity of the building, the enhancement of its longitudinal character, the simulation of a non-existent symmetry, the reinterpretation of its initial design, the reintegration of original elements, recovering the vernacular essence of a farmhouse and endowing, subtly, the building of new spatial values.

Mont-Ras Winery

The construction of a winery and the wine process creation are extremely attached at the experience with the land. The wine is smell, color, flavor and shape. The senses and the perceptions have to go together with a site that is able to emphasize the process of the transformation. In order of that, architects have worked with four key aspects:

The winery program is the result of the necessity to produce wine and organize a relationship between the existing land house. For the wine production there are four main spaces with three other ones between them, these last ones are the services spaces with all the facilities and storage. The first main space from the right, next to the laboratories and freezers, is the space for all the farming instruments and tools for the vineyards. The second one is for all the vats needed for the "mosto wine" production. The third one is for those vats and bottles that are resting. The last one, and forth, is the area for the Tastings, enjoyment and storage of the bottles that are ready to be open. One access through a tunnel from the upper side of the House is the one that organize the circulation of the owners. The access to the three other spaces is done directly from the vineyards.

The soil humidity helps the conservation of the wine. Architects decided to grave the winery into the earth to keep it with the ideal temperature and to create a platform for the existing House Land as well. The same earth is the one that helps them to create space.

The space deepness is the sound abortion, emptiness and shadow. The light organizes the space.

The building is a platform inside the earth. Its roof is a garden that lies on top of the concrete volts which its optimized calculations have drawn a section of hyperbolic arches. At the same time the platform is the water keeper for its re-use. The external walls are designed with the ideal shape having in consideration its material (the brick) to send all the efforts to the structure.

Architects: Jorge Vidal and Víctor Rahola / Location: Mont-Ras, Girona, Catalonia / Area: 573 m² / Photography: José Hevia

Pacherhof

The project is located in the isarco valley near the novacella abbey built in the 12th century. Here the monks used to produce wine from the nearby fields. Later many farms started their own wine business. pacherhof was the first but some documents show that their wine cellar was working before the construction of the monastery. This is why pacherhof is now classified as an asset of high architectural and cultural value.

From the historic cellar, through a staircase and a tunnel, you reach the new trapezoidal-shaped cellar below the existing land. In the highest corner of the plot emerges a pyramidal tower, clad in bronze panels that become part of the landscape contrasting with the peaks of the mountains. The tower houses an office and a tasting room on the upper floor. From here the winemaker can enjoy a view that embraces the vineyards, the old farm and the surrounding landscape. On the first floor, the bunches of grapes are harvested and then transported to the basement through an opening in the floor. In the basement, the production takes place.

The openings of the tower have no frame or shading, but are flush with the bronze panels, while the glass is treated with a bronzed effect. the aim is to create a solid, pure, monolithic geometric volume. The entrance to the cellar is marked by a concrete wall which has two functions: on the one hand it serves to direct the visitor towards the car park, and on the other it accompanies the ramp leading to the new cellar.

The choice of materials reinforces the contrast between the old and the new cellar: on the one hand the raw plaster and the metal for the extension on the other hand the smooth plaster and wood for the old cellar.

The old vaults and the new forms of expansion, the grey plaster of the existing cellar and the raw plaster of the new one establish relationships capable of highlighting the historical, cultural and sensory value of the intervention.

Architects: Bergmeisterwolf / Interior Design: Gerd Bergmeister, Michaela Wolf / Location: Italy / Photography: Gustav Willeit
Materials of Wine Rack: Wood, Metal, Stainless steel

The tower clad with bronze panels presents a funnel-shaped polygonal form.

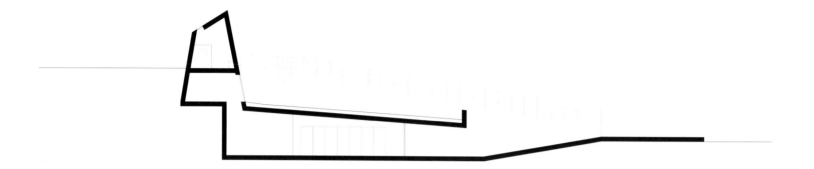

bergmeister**wolf**

SECTION   1:200

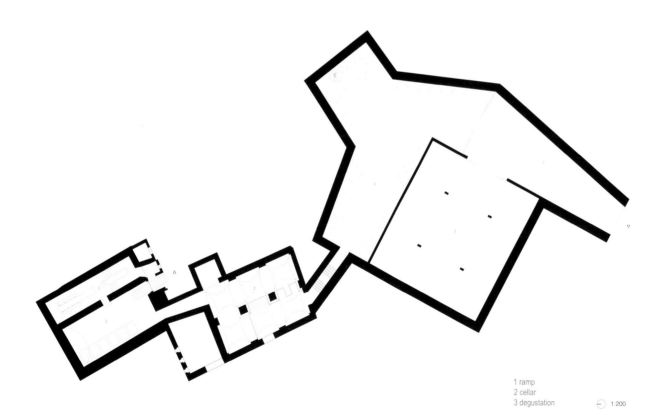

1 ramp
2 cellar
3 degustation   1:200

Details of the old cellar.

No treatment on the material, just to keep its original tones and make it age naturally.

Selected material marks the old wine cellar: rough plaster inside and metal for the extension.

A new metal staircase with a perforated pattern.

Production of wine, old cellar with wood barrels.

Production of wine, new extension cellar with metal barrels.

Loco Restaurant

The client specially requested to store a lot of wine in a visible way. He wanted the wine rack to be seen somehow. Since architects have thought about 3D tiles for the opposite wall, they thought, as well, why not create a sort of the same pattern for the wine storage area? And that was the principle behind the creation of this wine rack.

Instead of being just a wine rack, with no story and with not much interest, architects have made it a standout point of the entrance area, by using laser cut metal panels with the same pattern created by the 3D tiles in the restaurant's opposite wall. Something like a positive/negative effect, since the 3D tiles are standing out from the wall, and in the wine rack's case, the effect was made by cutting and removing material from the metal panels.

The shelves that support the wine were also laser cut made (but this time with a more simplified pattern, just the perimeter, giving perfect position to place the bottles of different wines and sizes in various positions. In order to make the effect even stronger, the whole wine rack was enlightened (led light was chosen since it is the source of light that hardly adds temperature).

Architects: João Tiago Aguiar / Design Team: João Tiago Aguiar, Renata Vieira, Ruben Mateus, André Silva, João Nery Morais, Ricardo Cruz / Location: Lisbon, Portugal / Area: 150 m² / Client: Perfume de Laranjeira, Lda. / Photography: FG+SG – FOTOGRAFIA DE ARQUITECTURA / Materials of Wine Rack: Glass, Led Lighting

A view of the wine rack when coming from the toilets.

Together with the hanging olive-tree, the wine rack is the ex-libris of the entrance.

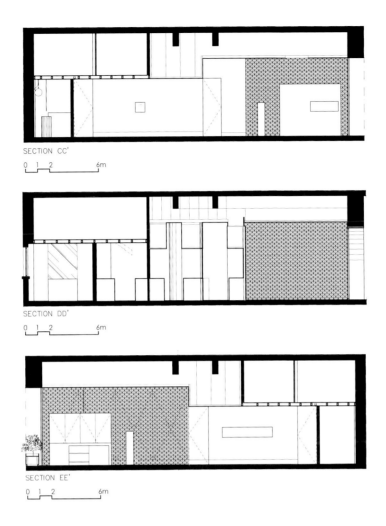

SECTION CC'

0 1 2   6m

SECTION DD'

0 1 2   6m

SECTION EE'

0 1 2   6m

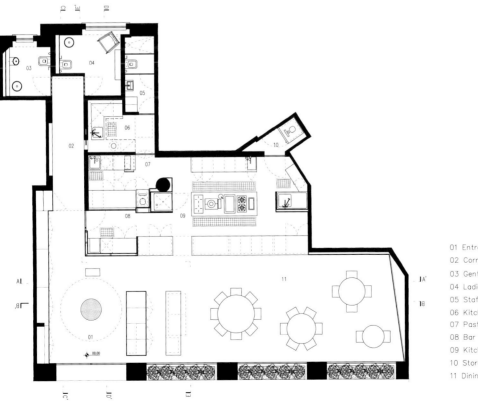

01 Entrance
02 Corridor
03 Gentleman's Bathroom
04 Ladie's Bathroom
05 Staff's Bathrooml
06 Kitchen's Unit
07 Pastry
08 Bar
09 Kitchen
10 Storeroom
11 Dining Room

PLAN

0 1 2   6m

The wine rack with the doors open. A more simplified structure then the one shown in its doors supports the bottles.

The pattern of the wine rack which resembles the 3D tile wall on the other side of the restaurant. Led light enlightens it all.

Mistral Iguatemi Wine Store

Seven years after the inauguration of its first store, located in JK Iguatemi Mall and With Arthur Casas' project, the distributor Mistral looked for Studio Arthur Casas team to sign its second address, now in Iguatemi Mall. Like the first, the new space should be inviting, innovative and surprising to provide customers with an enriching and enjoyable shopping experience.

Between stairs and accessed by to opposite entrances, the store allows customers to cross inside to reach parallel corridors of the mall. Taking advantage of this arrangement, Arthur Casas and the team created a path whose carbonized solid wood sides have shelves from floor to ceiling to accommodate the labels horizontally.

Remembering the old wineries, the carbonized solid wood wall was developed exclusively for this space thru a delicate process in which it's heated to a temperature of 3000 degrees Celsius to acquire special materiality: the brownish tone and a curvilinear character.

The sides of the path contrast with the bright off-white floor and ceiling. Such composition is accentuated by the lighting design, which makes walls look detached from the floor and the ceiling. Inside, the "Mesa amorfa" (amorphous table), also designed by the architect, serves as support for attending and for the wine exhibition. There is also a touch screen television for guests to delve into the history and beverage properties.

Located at one of the entrances, an air-conditioned wine cellar holds the finest labels and, in the other entrance, a window displays scenographically bottles and accessories on drawers, lit from the bottom up. To complete the program, a wine bar offers tastings and appetizers. The space, in the mall corridor, works in an island protected by slats of carbonized wood, following the same visual identity of the interior of the store.

Architects: Studio Arthur Casas / Design Team: Arthur Casas, Gabriel Ranieri, Nara Telles, Débora Cardoso, Raul Valadão / Location: São Paulo, Brazil / Area: 100 m² / Photography: Filippo Bamberghi

Mistral

Doluca Winery

The winery is situated in the middle of a planned industrialized zone on the outskirts of Çerkezköy and the scale of the building is an industrialized scale. As with all industrialized architecture utility, economy and logistics where key drivers. Yet unlike most factory settings, the "cultured" visitor is critical to the environment, especially in the harvest season. Employing landscape, altering scales, thresholds, light, color and materials, the site captures both the phenomena of the meta-context and of the craft of wine making.

Like the long history of utility buildings, passive is valuable for reducing operation costs, low-energy consumption and minimize logistics of non-localized materials/labor. The burning of earth from foundations to create immersive botanical gardens and contribute to the passive cooling of the barrack room, the use of skylights and penetrating light into deep floor depths and the prioritizing of all locally sourced materials were highlights of green design.

Like all factory buildings, the organizations that operate in them express their values and aspirations. Using sequence and drama we created a necklace of spaces the visitor experiences as well as is the path for everyday operations. Starting in the vast landscape of the industrial zone, one arrives to the warmth of the facade that changes its character with light through the day. The earth is shaped as the visitor gets closer and is eventually sculpted specifically to welcome the landscape into the building. Crossing a bridge that raises past the subtly sculpted earth, you are presented with morphology of the traditional doors of the village where the winery originated. With a slight recess of the threshold into the barrack room, one's curiosity is provoked as they are within the barrack room before they enter the building.

Architects: SANALarc / Location: Çerkezköy, Turkey / Area: 52,510 m² / Client: Doluca / Photography: Refik Anadol

şişeleme

tank alanı

1. Ana Giriş barlık alanından geçerek
2. Lobi
3. Veranda/Etkinlik
4. Konferans odası
5. Toplantı odası
6. Dinlenme
7. Laboratuar
8. Hazırlık odası
9. Tadım odası
10. Executive yemek odası
11. Yemekhane
12. Fabrikaya geçiş

1:200

1:500

ŞANAL ARC

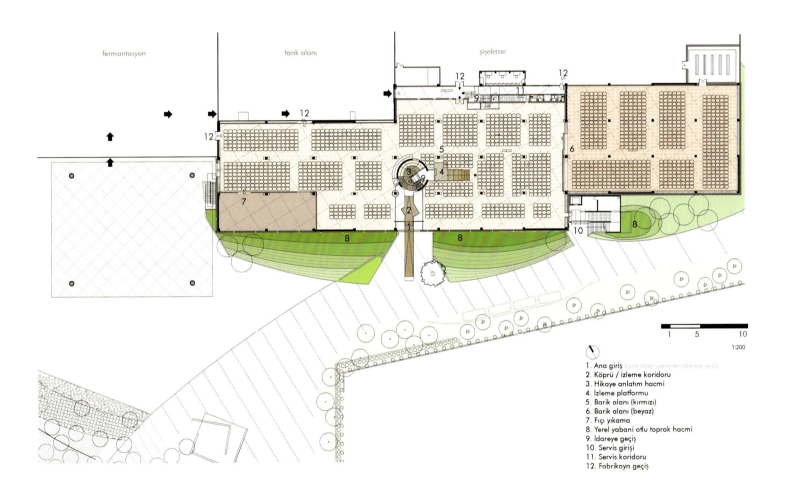

fermantasyon

tank alanı

şişeleme

1. Ana giriş *barik alanı üzerinden idareye geçiş*
2. Köprü / izleme koridoru
3. Hikaye anlatım hacmi
4. İzleme platformu
5. Barik alanı (kırmızı)
6. Barik alanı (beyaz)
7. Fıçı yıkama
8. Yerel yabani otlu toprak hacmi
9. İdareye geçiş
10. Servis girişi
11. Servis koridoru
12. Fabrikaya geçiş

1   5   10

1:200

Immediately upon entry the scale shifts along with the contrast of washed outdoor lighting into the lush moist quietness of the barrack room. The barrack's room's vastness is only slightly perceived as the visitor is now in a landscape of wood barrels. You are though reconnected back to nature and to the whole of Doluca's wine making organization with the Drum's central space. The arch, same as the entry into the barrack, is projected onto the vertical volume of light. The light is oriented to the sun to capture the changing seasons and times' of day; and bring the beauty of color and ambience into the heart of the facility.

One enters into the Drum with apertures to all of Doluca's core organizational functions. Moving along the edge of the volume you view the barrack room, the offices, the social spaces, the laboratory and the corporate lobby. At the end of the journey you are reoriented back into the industrialized landscape, but also the distant hills and natural landscape surrounding Çerkezkoy.

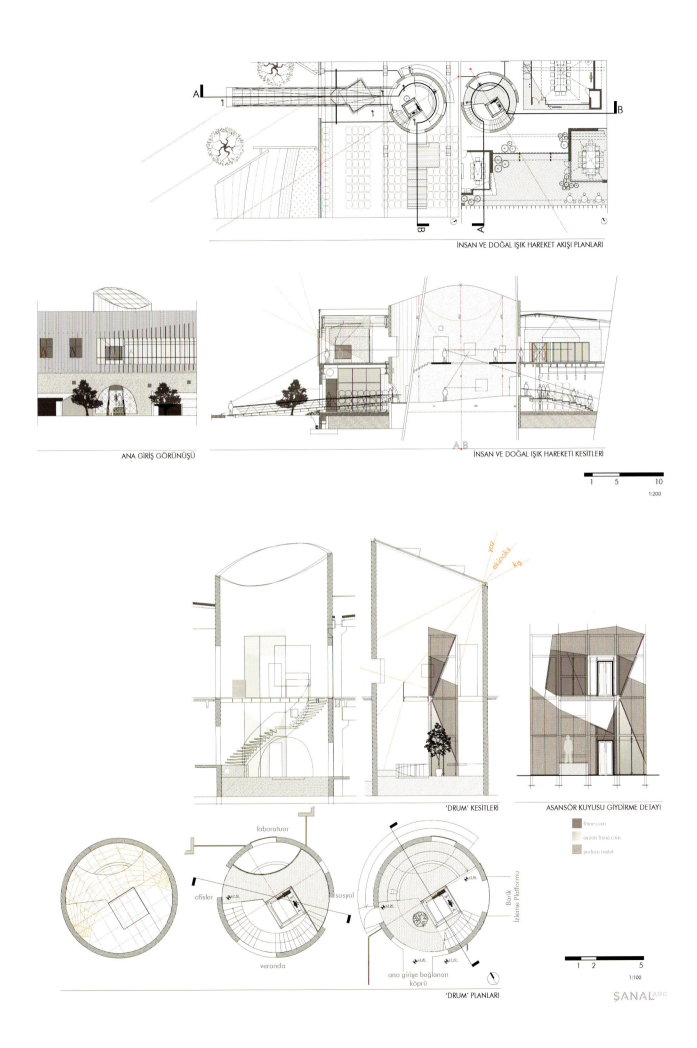

İNSAN VE DOĞAL IŞIK HAREKET AKIŞI PLANLARI

ANA GİRİŞ GÖRÜNÜŞÜ

İNSAN VE DOĞAL IŞIK HAREKETİ KESİTLERİ

1　5　10
1:200

'DRUM' KESİTLERİ

ASANSÖR KUYUSU GİYDİRME DETAYI

laboratuar

ofisler　sosyal

veranda

ana girişe bağlanan köprü

'DRUM' PLANLARI

1　2　5
1:100

ŞANALᴬᴿᶜ

Vegamar Seleccion Wine Shop

Wine shop Vegamar Selección, located on the main shopping street of town, is a space for wine tasting and sales. The project aims to transmit the quality of the exposed products, as well as to maximize the feeling of amplitude inside the shop.

Dark, glossy panels are chosen for vertical surfaces. Due to their tone and reflections, they blur the spatial limits of the establishment, making it seem bigger than what it really is. The colour of this material also refers to the colour of the displayed wines. These vertical surfaces lodge storage space and allow for a regularization of the available geometry. The back of the wine bar contains backlit furniture which — along with the upper mirror — optically doubles the perception of space.

The first area is used for product exhibition. Two horizontal rips in which wines coming from the Vegamar wine cellar are presented. These elements along with indirect lighting on the ceiling, indicate towards the inside of the shop. At the end of the tour — in an area doubling the original width — is where the wine tasting takes place.

This is a project that seeks to bring into the city the work done in the wine cellar.

Architects: Fran Silvestre Arquitectos / Interior Design: Alfaro Hofmann / Project Team: Fran Silvestre, Jordi Martínez, Ángel Fito
Location: Valencia, Spain / Area: 123 m² / Photography: Diego Opazo

VEGAMAR
SELECCIÓN

VEGAMAR
SELECCIÓN

De Lunes a Sabado
de 10 a 22 horas

Electric roller shutter door, colour matt black; Sunshade, colour matt black.

P01. TIENDA - EXPOSICIÓN SHOP - EXHIBITION   32,00 m
P02. ESCAPARATE - MOSTRADOR WINDOW DISPLAY - COUNTER   6,50 m
P03. ESPACIO DE DEGUSTACIÓN DE VINO WINE TASTING SPACE 48,40 m
P04. COCINA KITCHEN   10,55 m
P05. ASEOS BATHROOM   12,20 m
P06. DESPACHO OFFICE   14,15 m

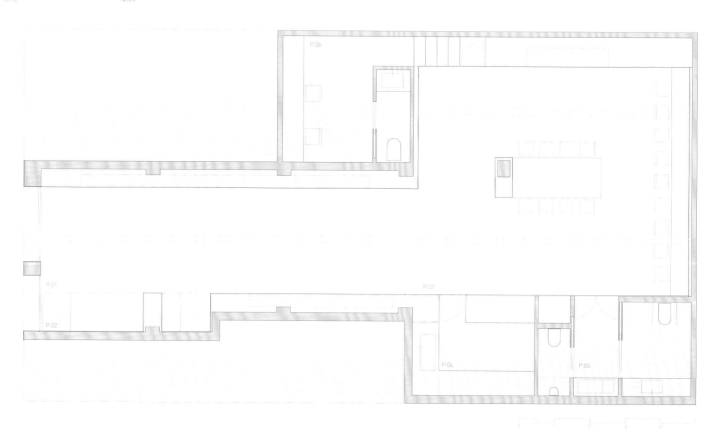

Laminated parquet flooring, colour grey with acoustic insulation.

Black lacquered MDF panels (high-gloss).

Table with MDF board high gloss lacquered (19 mm), colour matt black stool.

Weinverkauf Eisacktaler Kellerei

The original winery complex was built in the 1970s in the industrial area of Chiusa. Initially, the building was designed as an industrial shed, highlighting this character in the appearance of the facade. In order to deconcentrate internal work and improve quality, it was decided to demolish part of the building and replace it with new buildings.

The new building is an entirely concrete structure without windows. The external cladding is made of 12 cm thick, grey-red plaster. All the steel elements, such as the roofs and doors, are in the same shades of brick red. These chromatic choices guarantee a total harmony between the building and the natural context in which the project is inserted, in the same way the choice of facade recalls the nearby rocky formations.

The cellar of the barrique has a ceiling divided into small barrels, covered in brick, while the walls are made with a plaster based on clay. The interiors are therefore dominated by red tones and burnt earth, colours that recall the tradition of whine.

Architects: Markus Scherer / Collaborators: Marion Trafoier, Miriam Lopez, Elena Casati / Location: Klausen / Client: Eisacktaler Kellerei / Area: 260 m³ / Photography: Georg Hofer (Miriam Lopez)

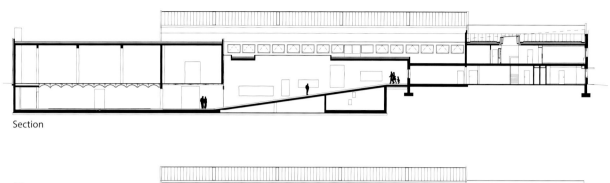

Section

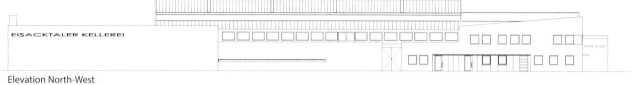

Elevation North-West

Elevation South-West

Eisacktaler Kellerei (Klausen) / Cantina valle Isarco (Chiusa) -   Section and Elevations

0 1 2   5   10

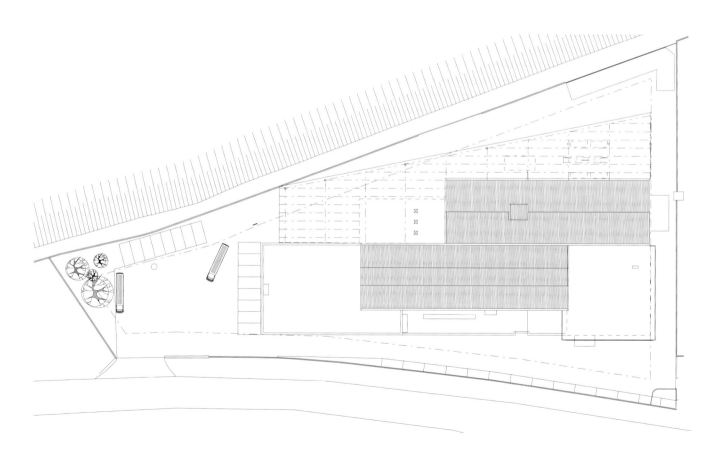

Eisacktaler Kellerei (Klausen) / Cantina valle Isarco (Chiusa) - Site plan

0 1 2   5   10

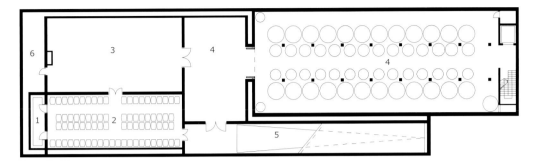

Basement Floor

1 bottle storage
2 barrique cellar
3 wooden wine barrels
4 storage
5 ramp
6 technical area

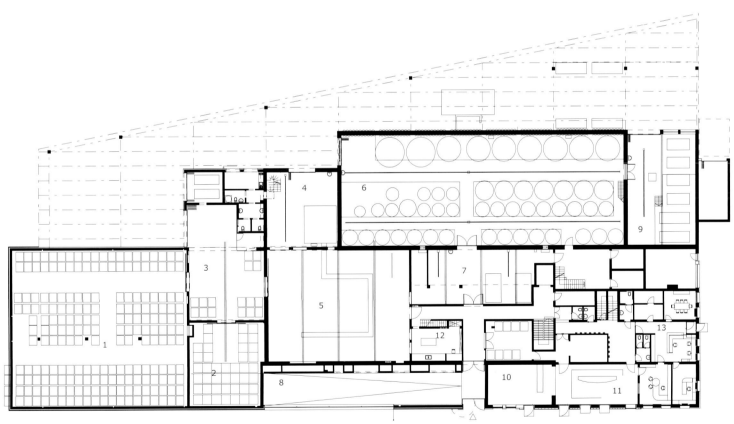

Ground Floor

1 cold room
2 cartoning
3 delivery
4 washroom
5 botting room
6 fermentation room
7 engine room
8 ramp
9 winepress room
10 bar
11 shop
12 laboratory

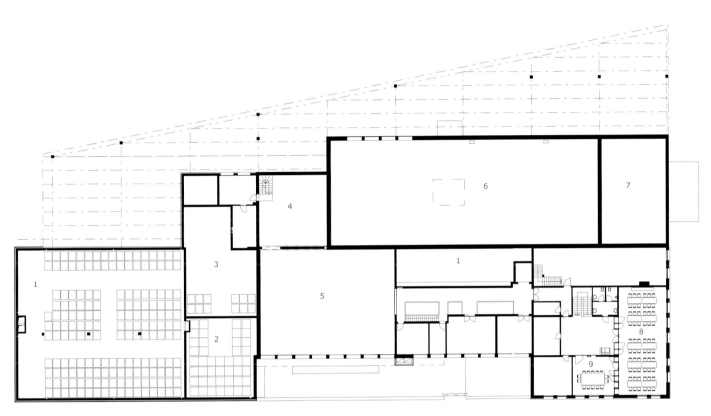

First floor

1 cold room
2 cartoning
3 delivery
4 washroom
5 botting room
6 fermentation room
7 winepress room
8 room
9 conference room

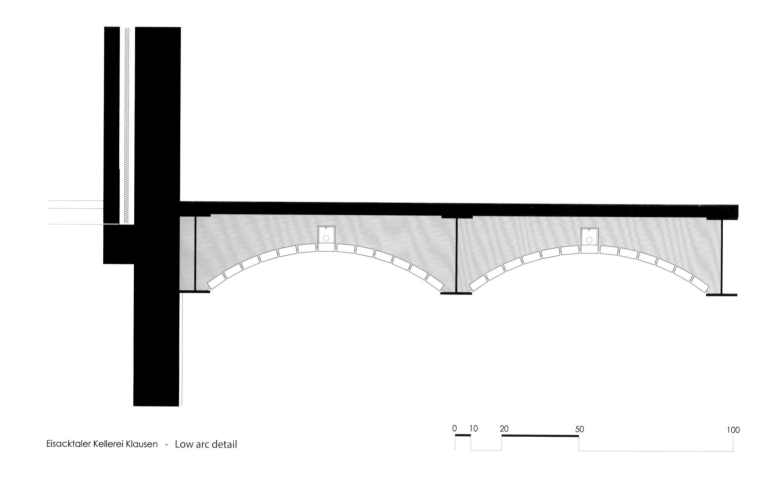

Eisacktaler Kellerei Klausen - Low arc detail

Wine Cellar in Slovenia

The small vine cellar is situated in the basement of the existing holiday house in the wine region of the southeast part of Slovenia. It was used as a general storage space for the old barrels and other amenities. During the reconstruction of the holiday house in 2007, the storage was pierced through with two concrete columns and beams to provide necessary structural reinforcement. The refurbishment, executed several years later, took these concrete elements as a main conceptual start-up.

The columns now serve as a support for the cantilevered concrete bars, refined and polished in a similar manner. Floor follows the same materiality, while the rest of the surfaces and the furniture is made out of oak and birch wood. The general exception and specialty of this project are the 200 concrete tubes stacked into a wooden wall. As the wine cellars nowadays serves more as a place for celebration and socializing, the memory of the bottles, used to be filled with precious wine, is now petrified and replaced in another material.

The use of the ambient light, positioned in the void spaces between the concrete bottles, enhance the created meaning of the interior design on one side, and reduce the overall weight of the elements on the other. The careful balance between the domestic (and almost vernacular) use of wood and the austerity of the concrete elements manages to position this renovation into the dominating traditional landscape and general cultural milieu with necessary freshness and ambiguity.

Architects: Studio Abiro / Interior Design: Studio Abiro / Design Team: Dr. Matej Blenkuš, Katja Cimperman, Anja Cvetrežnik
Location: Brežice, Slovenia / Area: 23 m² / Photography: Miran Kambič / Materials of Wine Rack: Wood, Concrete

Even though the space is relatively small, there are few distinctive atmospheres there, constructed by variety of the surrounding surfaces. Yet, only two materials are used – wood and concrete.

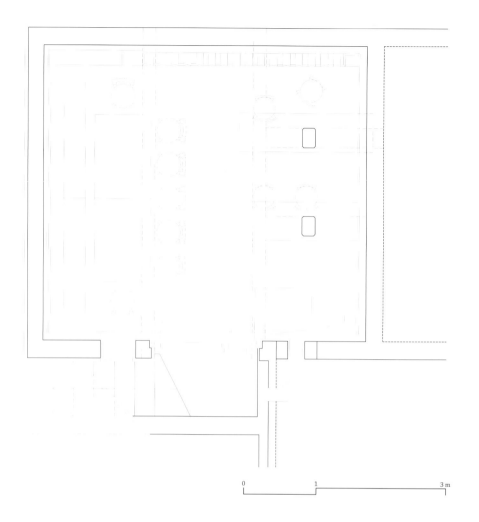

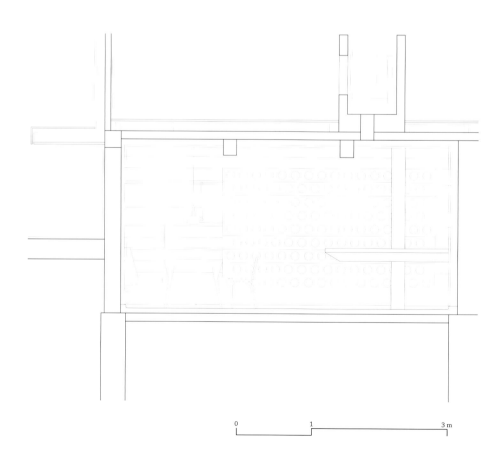

The main goal of the wine cellar is providing a pleasant space for smaller or larger groups of guests. The gathering is organized around big table and two concrete bars with stools.

Two concrete columns and ceiling beams are the exiting structural reinforcements added to the wine cellar during the major reconstruction of the building. They have been taken as an existing starting point of the project, defining the position of the concrete bars and the overall materiality.

The surfaces of wall and furniture are made out of oak and birch wood. The selection of darker wood for wall framing and lighter for the furniture creates modest spatial contrast which emphasizes the content.

Generally wine cellars are defined by the set of bottles. As the climate for storing wine is not suitable to be mixed with well-being of people, the challenge is to recreate the spirit of the wine cellar – without actual bottles of wine.

Espai Saó

The Espai Saó, together with the Gallery House, is part of a series of interventions carried out for the Mas Blanch i Jové winery, located in the small village of La Pobla de Cérvoles, in Lérida. This new space, built inside the large wine production hall, responds to the growing demand for events that are being generated in the winery, for which the previous tasting room had remained small.

The position was delimited beforehand by the available space in the production hall and by the necessary connection with the previous tasting room, resulting in a rectangle in plan of twenty one meters long and six meters wide, covering the entire width of the hall, and raised above the working area as a mezzanine. The connection with the previous tasting room is made by means of a bridge, of generous dimensions, which hangs from a pair of cables on the room of barrels, to serve as viewpoint of the same one, carried out by a very extensive mural of Gregorio Iglesias that covers all the walls of the room.

The flexibility required for the space was to function as a tasting and events room, restaurant, and exhibition hall. This flexibility translates into a division in the cross section of the room: a strip of 1.60 meters wide runs through the room from end to end, illuminated by three skylights, occupying all the available free height, and covering its walls in white, designed to serve as an exhibition space; a second strip of 4.80 meters wide is occupied by the events and tastings room, on which a spacious false ceiling develops where all the machines, ducts, lighting… run.

Architects: Raúl Sánchez Architects / Design Team: Miriam Corcuera, Valentina Barberio, Albert Montilla / Location: Mas Blanch i Jové Winery, La Pobla de Cérvoles (Lérida) / Area: 120 m² / Client: Mas Blanch i Jové / Photography: José Hevia

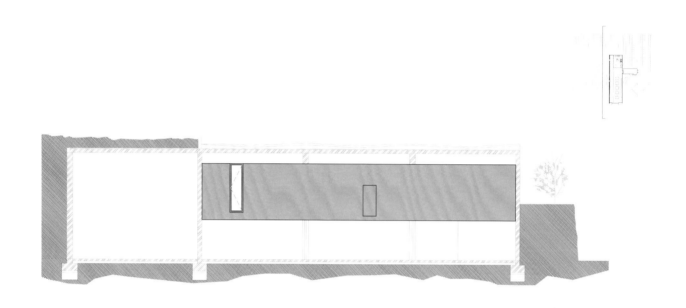

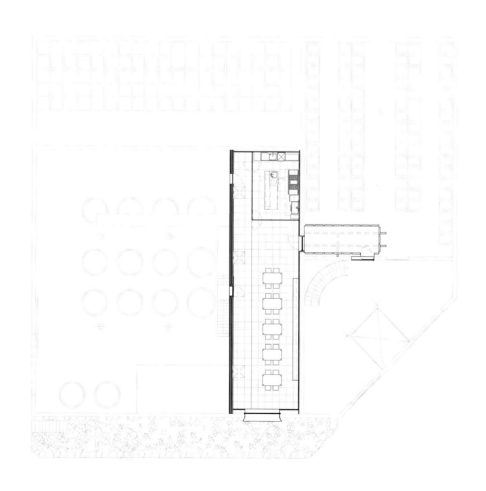

Everything is a single space, except for the mandatory enclosure of the
kitchen, but the two main uses are defined by a dialectical play of textures
and materials: the event zone is covered in walls and ceilings of mass colored
MDF boards, in black color, modulated with open joints covered by corten steel plates that establish a veiled relationship with the same material
used by Guinovart on the other side of the wall in the former tasting room;
the exhibition area is covered in white. The flooring is the same, but a slight
change of color marks a line of separation between both stripes, aligned with
the height jump of the roof.

On the other side of the room, a hollow was cut out on the concrete facade
wall, destined to show the impressive landscape, as a living canvas, framed in
cor-ten steel plates. The furniture, consisting of tables and chairs all made of
solid oak, offers a natural counterpoint to an interior that is intended to be
cooler.

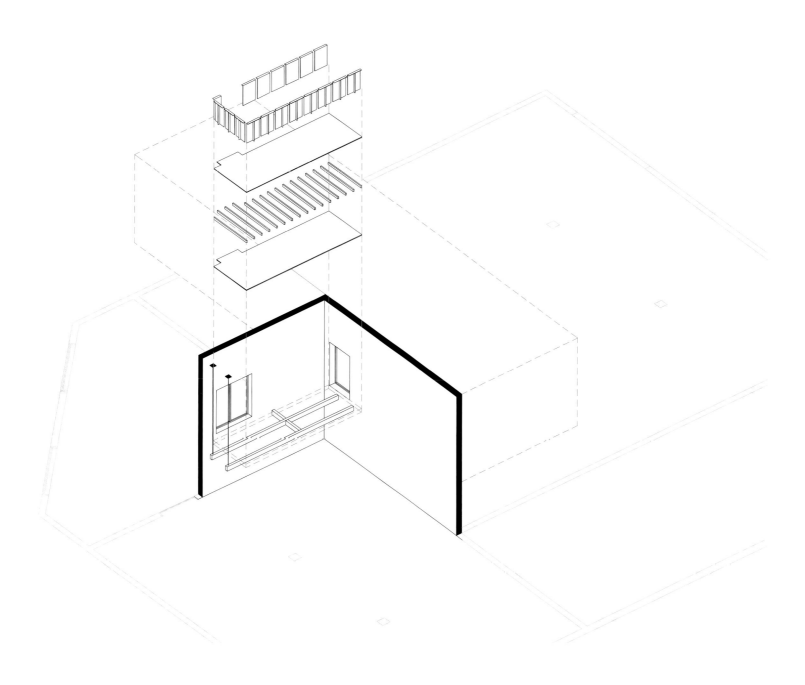

Kutjevo Winery

The winery in Kutjevo lays in the middle of continental Croatia's region of Slavonia, Vallis Aurea, the cradle of winemaking in Croatia, where the Cistercian monks in the early thirteenth century founded the first abbey together with a wine cellar, and started producing wine.

Although many other prominent and awarded winemakers live and operate nowadays in Vallis Aurea, still it remains underdeveloped, which makes any business initiative an important one, for the whole region. By building a winery, basic conditions are created for the following: the development of new technologies in the production of wine, enlargement of the existing capacities in vineyards, modernisation of the touristic presentation techniques, new employment opportunities, and so on, all in the service of establishing a new set of standards concerning wine tourism and the following industry.

The winery is placed in the Kutjevo city centre, on the main street near the castle (original abbey), on the spot where an old school used to be. Although it is a city centre, the houses are very modest which is why architects wanted the winery to be primarily experienced as a city house. Only upon entering, the visitor would discover it is a production facility as well.

A clear and basic architectonic design derives from the intent that the house fits into the city structure and yet remains contemporary. The initial inspiration comes from the traditional wine cellars and country houses. Buildings of the area often have concrete bases that are unfinished and brick walls which gain in patina and robustness in time. The winery itself is divided into two parts; the base volume is in concrete vaults which both open and hide the space of wine ageing barrels.

Architects: Dva arhitekta / Location: Kutjevo, Croatia / Area: 3330 m² / Client: Vina Galić / Photography: Damir Fabijanic

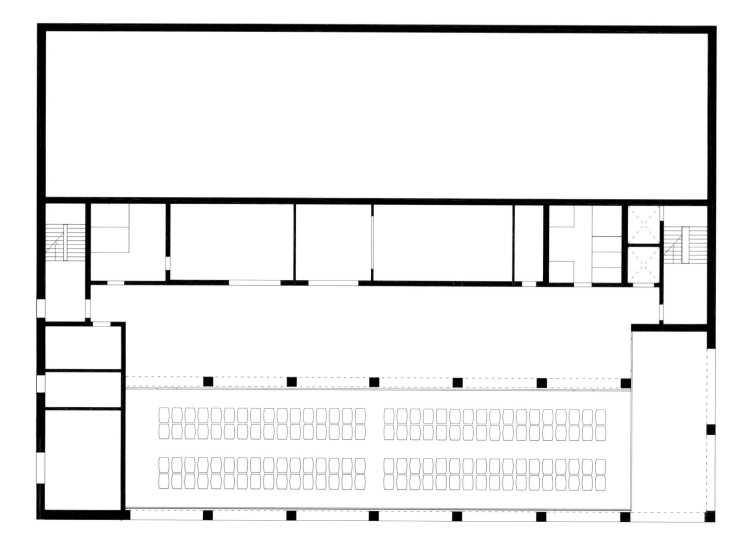

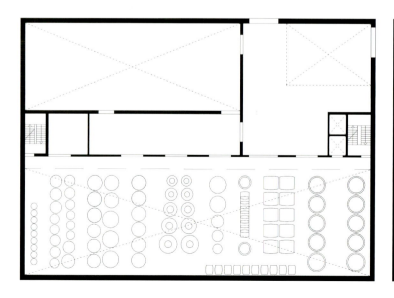
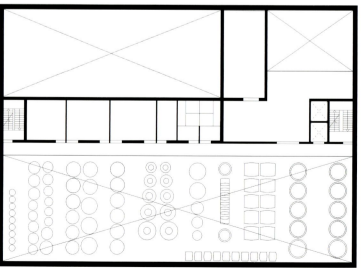

Contrary to the usual winery scheme where the barrels are below the ground in a cellar, here the barrels are placed into street window, to create contact with the street and invite passer-bys. Concrete base also comprises wine degustation facilities, spaces to experience the wine. The upper part of the winery cladded in brick houses the wine production and the wine storage, which communicate with the city via back street and courtyard.

The owner decided to invest in high end wine producing equipment (mostly custom-made), which define the modern interior of the winery, together with visible concrete on the walls, terazzo flooring, and details in wood.

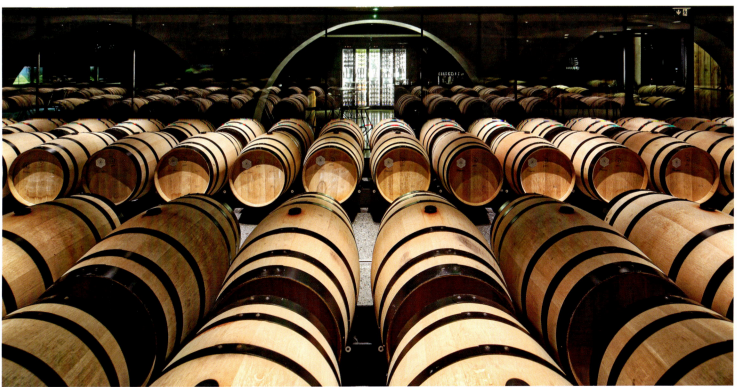

Segura Viudas – Eco Winery

INDAStudio designed the public areas for the renovation and restyling of Bodega Segura Viudas. Located in the heart of the Penedès, prestigious vineyard territory of Catalunya, Segura Viudas is home to producers of the traditional sparkling Cava and other boutique wines. The project has been realized within the ancient eleventh-century farmhouse which originally served as a defence tower to domains belonging to the Monastery of Sant Cugat in Barcelona.

The interior design was inspired by restyling of the Segura Viudas branding in combination with the property's original architecture and morphology. The renovation enhances several spaces in the property but a real eye-catcher is the new hall. Here visitors are welcomed to the cellars with wine samples and tapas. Members of exclusive wine tasting clubs can enjoy a separate reserved area the most distinguished Cavas and wines from Bodega Segura Viudas. This space is divided by an impressive Ogival arch that separates the lobby from the wine tasting area.

Romanesque structural elements of the property, such as the 'masonry' walls made from irregular cut stones, the arches (semicircular and Ogival) and the 'loophole' windows have all been partially restored. A unique paper lamp designed by Ingo Maurer floats in the middle of the room, above the new tasting area. Maurer's lamp illuminates the furniture with a rustic but contemporary style, blending perfectly with the orange tones of the walls. A sample of the complete product collection of Bodega Segura Viudas is displayed in a classic wooden bookcase alongside the awards given for their most prestigious Cavas and wines. The rustic atmosphere is completed by natural fabric carpets which partly cover the original floor.

Interior Design: INDAStudio / Design Team: Isa Rodríguez, Silvia Ros / Location: Barcelona, Spain / Area: 500m² / Client: Heredad Segura Viudas / Photography: Mercè Gost and Freixenet Group / Materials of Wine Rack: Wood, Metal, Wrought iron

Segura
Viudas

View from the lobby of the wine tasting area.

General floor plan

1.   Outdoor Square
2.   Ancient Farmhouse Building
3.   Old Vault Cellar
4.   Outdoor Cellar
5.   Wine Bottling Line & Casks Storage
6.   Sparkling Wine Warehouse
7.   Wine Warehouse
8.   Purifying Plant

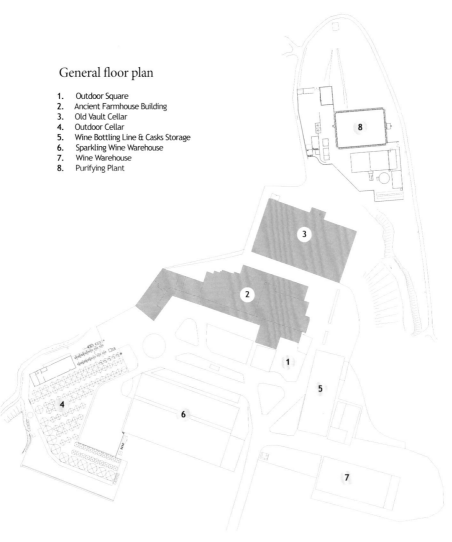

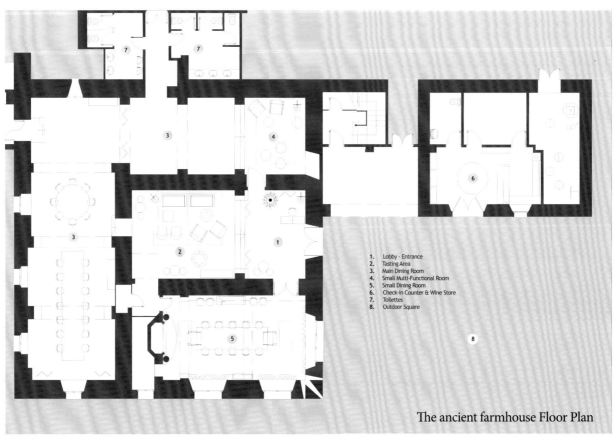

1.   Lobby - Entrance
2.   Tasting Area
3.   Main Dining Room
4.   Small Multi-Functional Room
5.   Small Dining Room
6.   Check-in Counter & Wine Store
7.   Toilettes
8.   Outdoor Square

The ancient farmhouse Floor Plan

Frontal view of a wooden bookcase.

The main dining room has a capacity of 60 people and a 125 m² surface. Linked to the hall by one of the property's many semi-circular arches, the space is split in two by an L-shaped floor plan. The room features a new lighting system, comfortable furniture and Ogival arches holding a ceramic vault and wooden beamed ceiling. At the end of the dining room, there is a small multi-functional room that can be used for tastings, projections or networking. Special attention was given again to authentic details such as the preservation of the walls' original medieval stones and loophole windows.

The old bodega has been kept in its original state, preserving the ancient architecture and the wine-storage facilities of 50 years ago. During a visit you can see the bottles displayed in 'rime', that is, one above the other in a horizontal position. They cover the side walls forming enormous compact blocks. This grouping occurs due to the production of Cava by the traditional method in which a second fermentation takes place while in storage. Visitors can also common see the classic wooden shelves that are used in the manual removal stage of the bottles, another fascinating phase of the elaborate Cava making process after the 'rime'. All this producing ritual, in combination with the stone, light effects and unique history make a visit to Bodega Segura Viudas unforgettable.

The wine store and small dining room have also been renovated with new branding according to the concept of Sustainable Eco-Winery Boutiques. For this purpose, walls and ceilings were treated in soft white colours leaving the wooden beams in their original state. Existing architectural elements have been restored, the furniture and lighting system renovated and decoration added to create harmony with the new vision of the property. Materials such as wood, wicker, glass, ceramics, handmade macrame tapestries and other natural finishing materials were a source of inspiration to complete the project.

Products displayed in bookshelves in the wine store.

View of the main dining room.

Wooden shelves for storing champagne bottles during the fermentation process.

Stacked bottles covering the walls of the old vault cellar.

Colono Shop

The project tries to update the concept of the traditional grocery store, adapting it to new and different situations. A black wood perimeter is proposed, operating as a wrap housing the facilities and large appliances, together with a central piece of furniture, made of pine wood that colonizes the different spaces in which the franchise can be located.

Taking the fourteenth-century "portulano" maps as a reference and abstracting the geometry of their directional lines, architects generated four pieces of wood with different angles that, combined in a myriad of possibilities, give rise to a series of modules designed to host various activities. This modular system allows the creation of multiple custom geometric configurations according to the particularities of the premises to be intervened or the convenience of establishing certain uses to a greater or lesser extent according to regulations, location, amount of initial investment or marketing strategies. They have common heights, geometries and material, so a homogenous formal identity is always maintained, being Colono recognizable anywhere in the world.

This flexibility, based on 4 unique generating elements, means that, regardless of the use of each module, they all have an exhibition surface on both sides with a contrasting background, being the products a backdrop for any activity to be carried out. The furniture, which is presented as exempt, colonizes the space accommodating its position and shape to the perimeter in which it is inserted, and illuminating both the exposed product and the space, generating an intimate and welcoming atmosphere.

Interior Design: Serrano + Baquero Architects / Design Team: Paloma Baquero Masats, Juan Antonio Serrano García
Location: Wien, Austria / Photography: Fernando Alda / Materials of Wine Rack: Wood

Details of Colono furniture, with different heights and geometry.

View of Colono furniture in the entrance, transforming itself to be bar and product preparation area.

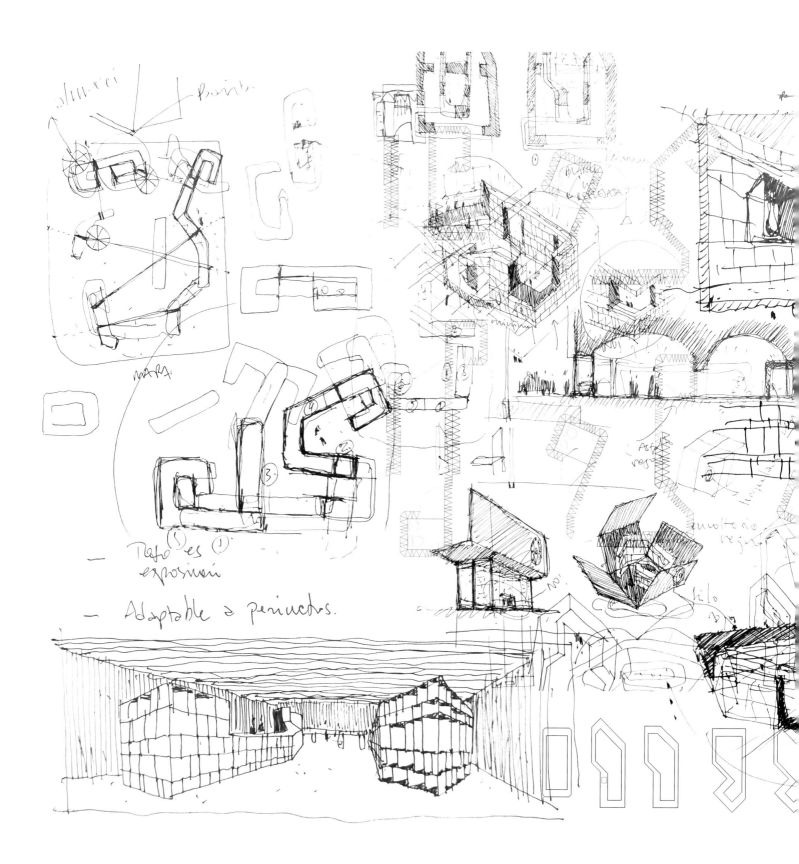

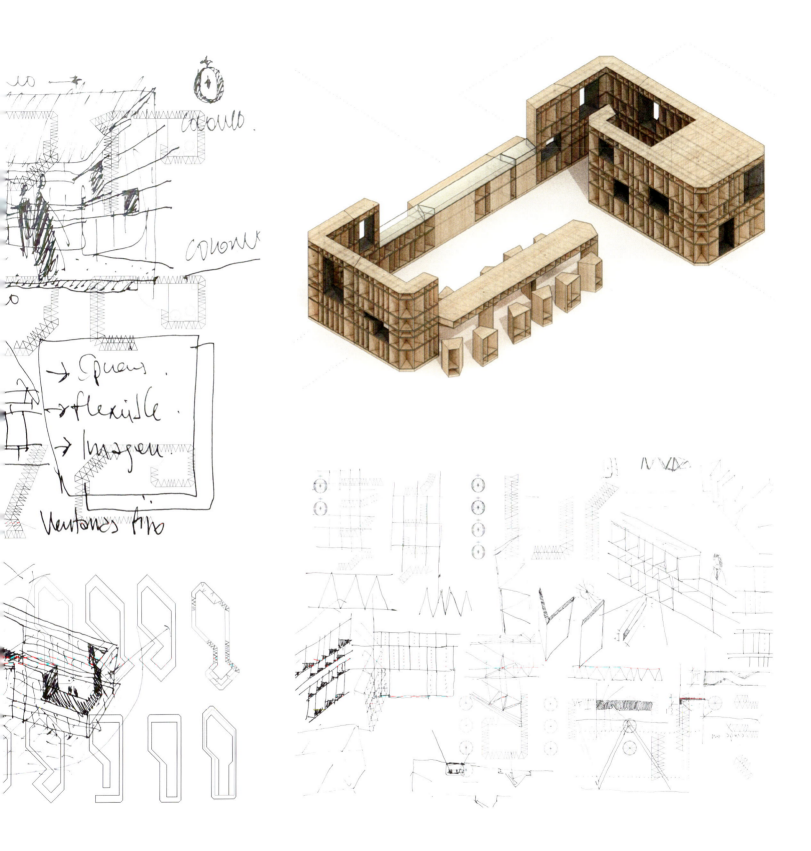

Tannic by Freixenet

INDAStudio designed the first showroom for the famous Spanish sparkling wine company in Barcelona baptized with the name Tannic by Freixenet. The concept of the project was inspired by fusing a traditional wine cellar atmosphere with a modern urban space. Noble materials such as natural wood, black iron, white marble and glass were used to design the furniture and the wallcoverings.

The space has two floors each with different functionalities. The showroom is located on the ground floor. The walls and the ceiling are partially covered with natural oak — a typical wood used for making wine barrels. The phrase 'The Ferrer family wines' is engraved in the wood in fifteen different languages to reflect the presence of the company and its products around the world.

An angular marble bar also serves as a counter space, cash register and tasting area. Globe crystal lamps that allude to the form of a grape, hang up above the bar. Their colours evoke the two leading products of the company: the gold lamps speak of the sparkling wine or "cava" and the burgundy lamps invoke the image of red wine.

The products are displayed in elegant black iron shelves, one of which acts to visually connect the ground floor with the first floor. Pieces of wood and boxes were added to the composition giving warmth to the shelving unit and incorporating the element of traditional wine boxes. Black iron was also used in selected areas as a wall covering and to design other decorative and functional elements like counterweights and signage.

Interior Design: INDAStudio / Architects: The Floor Bcn / Design Team: Isa Rodríguez, José Luís Bonet, Bel Diví / Location: Barcelona, Spain / Area: 110 m² / Client: Freixenet Group / Photography: Freixenet Group / Materials of Wine Rack: Wood, Metal, White Marble, Glass, Black Iron

View from the entrance of the showroom and counter space.

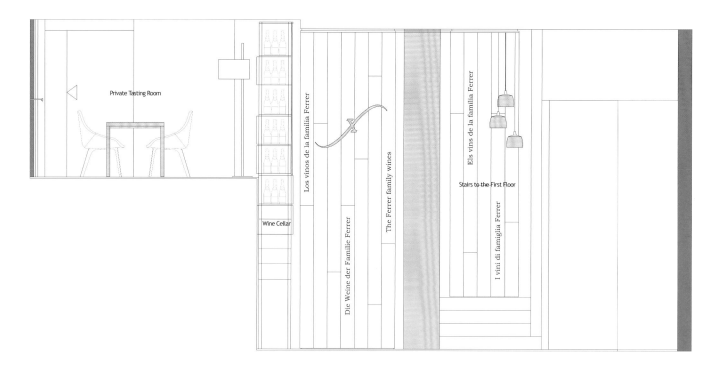

Cross Section

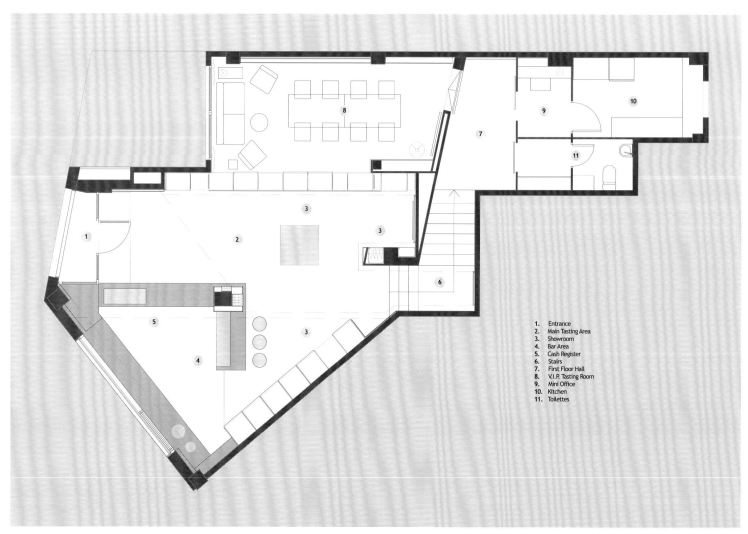

1. Entrance
2. Main Tasting Area
3. Showroom
4. Bar Area
5. Cash Register
6. Stairs
7. First Floor Hall
8. V.I.P. Tasting Room
9. Mini Office
10. Kitchen
11. Toilettes

General Floor Plan

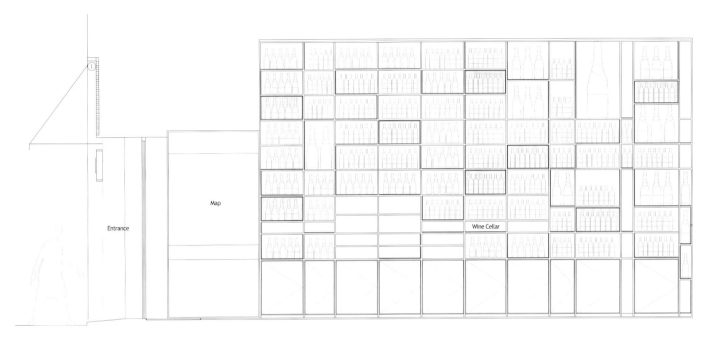

Long Section

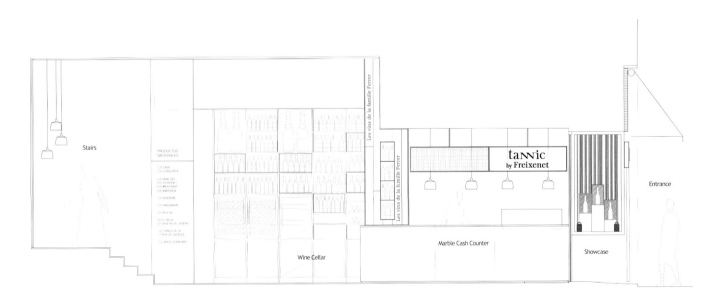

Longitudinal section

The angular marble bar and one of the black iron shelves.

A private tasting room, a small office and the toilettes are on the first floor. Part of the showroom can be viewed from the tasting room through the iron shelving unit that connects the two levels. A wide horizontal mirror dominates one wall which makes the room feel more spacious. A long oak table stands in the centre of the room and is used as the principal furniture for the tastings.

Special guests and VIPs can enjoy the quality products of Freixenet sitting in comfortable sofas and armchairs that complete the decoration and functionality of the space.

Hilton Adelaide

Global design studio Landini Associates has again teamed up with Hilton Hotels, this time in Adelaide, to design their new restaurant, 'Coal Cellar & Grill' and adjacent bar and hotel lobby lounge area. Simply, the brief was to design a restaurant that did not feel like a hotel, a restaurant that would be a destination in and of itself for the local community.

For 35 years the restaurant was run by the Penfolds winery creating a strong heritage but a hard act to follow. Since its closure, the space reverted back to a hotel offers with little character. The brief was to reinvent a destination for the city centre and a sanctuary for the hotel's guests.

Planning was a key factor in Landini Associates' design solution. The bar services both the lobby lounge and the restaurant, acting as a fulcrum between the two spaces. At one end is a floor to ceiling wine room, housing over 3000 bottles; at the other a suspended charcuterie and cheese display. Both design features are visible from all angles of the restaurant and hotel lobby.

Within the restaurant itself, planning was also crucial. The nature of hotels meant the restaurant had to be suitable for all-day dining, from breakfast through to the evening, and cater to a range of covers. Four private dining rooms have been created north and south of the main restaurant, designed with large scale sliding doors that can be opened up as an extension of the restaurant, or closed off to ensure that the restaurant does not feel empty in quieter times. They can be used as intimate dining spaces for larger groups, or conference rooms. The open kitchen and grill provide a backdrop of theatre, enhancing the dining experience.

Lighting throughout the whole space was consciously designed with five different settings, based on creating the right ambience from breakfast through to evening.

Architects: Landini Associates / Interior Design: Landini Associates / Location: South Australia Area: 1200 m² / Client: Hilton Hotels / Photography: Trevor Mein / Materials of Wine Rack: Stainless steel, Wrought iron, Glass

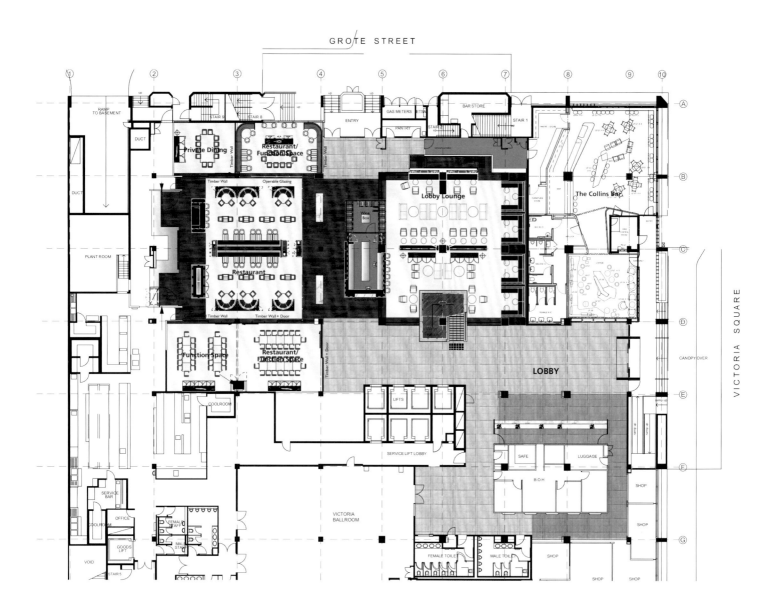

The material palette is a simple one. Dark timbers, leather, stone and glass form the basis of the backdrop complemented by an exposed ceiling and new lighting scheme. Exposed concrete columns and ceiling slabs speak a language of honesty whilst embracing the existing building bones. The important thing was to create a space that supported the restaurants new food offer which was coined "Seriously South Australian".

Cehegín Wine School

The former cellar where the project was developed belongs to a landmark building called "Casa de la Tercia", dated from the seventh-century. Located on the semi-basement floor of the building. The purpose of the intervention project is to enhance its value through the development of an oenology school, exhibition space and museum for its visitors.

Historically, the grapes were pressed, in the building's courtyard, the juice flowed into the cellar, travelling along channels carved into low-ceilinged limestone vaults, then into pine scuppers emptying into rows of a large urn in unglazed terra-cotta. The urns were partially sunken in the earth to keep the maturing vintage at cool temperatures until it was fully aged.

According to the project, we turn this interesting space into a gallery, with 342 m^2, where the main vault from the entrance is crossed over by a glass ramp that slopes gently down before levelling out to become a carpet of glass, a walkway that is transparent and colourless but full of fleeting reflections. At the far end, an open space hosts lectures and concerts, while the kitchen supports catered functions and cooking classes with wine pairings.

Crossing a wall we find the Wine shop, with interesting details, for instance, the necks of the bottles gripped by built-in pine yokes. At the bottom, we can find another small gallery and a kitchen. Here almost everything new is hidden as much as possible be. Conceptually, raw steel and wood are very elemental materials that work well with the historic parts of the interior, respecting the handmade character and the patina of time.

Design Agency: INMAT Arquitectura / Location: Murcia, Spain / Area: 342.75 m² / Photography: David Frutos / Materials of Wine Rack: Wood, Metal, Stainless steel, Wrought iron, laminated glass

The old access to the cellar is closed by means of a sliding door composed of a metal structure and a glass sheet, adopting the same proportions as the upper fence.

The new circulation surface is embedded between the ancient walls and jars.

A small exhibitor next to the exit introduces us to the concept of the gallery.

The stairway to the upper floor building embraces the wine cellar and makes it part of the structure.

The wine cellar, created specifically for this space, is part of the store area.

Antinori Winery

The site is surrounded by the unique hills of Chianti, covered with vineyards, halfway between Florence and Siena. A cultured and illuminated customer has made it possible to pursue, through architecture, the enhancement of the landscape and the surroundings as expression of the cultural and social valence of the place where wine is produced. The functional aspects have therefore become an essential part of a design itinerary which centres on the geo-morphological experimentation of a building understood as the most authentic expression of a desired symbiosis and merger between anthropic culture, the work of man, his work environment and the natural environment.

The physical and intellectual construction of the winery pivots on the profound and deep-rooted ties with the land, a relationship which is so intense and suffered (also in terms of economic investment) as to make the architectural image conceal itself and blend into it. The purpose of the project has therefore been to merge the building and the rural landscape; the industrial complex appears to be a part of the latter thanks to the roof, which has been turned into a plot of farmland cultivated with vines, interrupted, along the contour lines, by two horizontal cuts which let light into the interior and provide those inside the building with a view of the landscape through the imaginary construction of a diorama.

The facade, to use an expression typical of buildings, therefore extends horizontally along the natural slope, paced by the rows of vines which, along with the earth, form its "roof cover". The openings or cuts discreetly reveal the underground interior: the office areas, organized like a belvedere above the barricade, and the areas where the wine is produced are arranged along the lower, and the bottling and storage areas along the upper.

Architects: Archea Associati / Location: Bargino, Florence, Italy / Photography: Pietro Savorelli

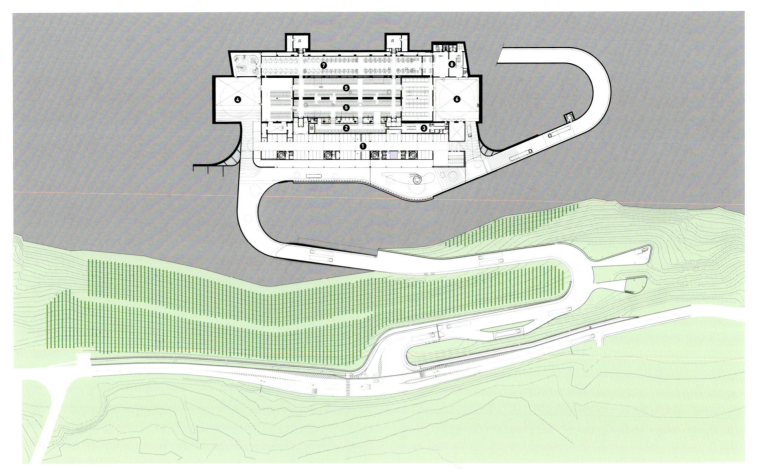

Pianta quota +169.65 / Plan level +169.65

1. Parcheggio bottaia / Bottle cellar parking lot
2. Bottaia / Bottel cellar
3. Riserva della casa / House reserve
4. Piazzale barriques / Barrique square
5. Barricaia / Barrique cellar
6. Piazzale tini / Vat square
7. Tinaia / Vat cellar
8. Camera presse / Wine press room

ARCHEA ASSOCIATI
CANTINA ANTINORI
ANTINORI WINERY

Pianta quota +169.65 / plan level +169.65

0   10   30   50 m

ARCHEA ASSOCIATI
CANTINA ANTINORI
ANTINORI WINERY

sezione GG / section GG

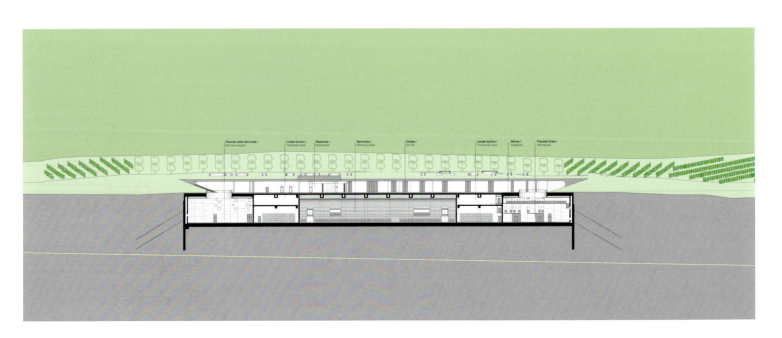

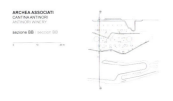

ARCHEA ASSOCIATI
CANTINA ANTINORI
ANTINORI WINERY

sezione BB / seccion BB

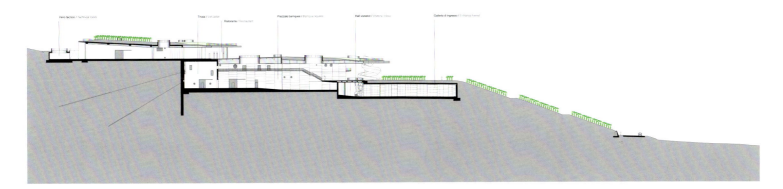

ARCHEA ASSOCIATI
CANTINA ANTINORI
ANTINORI WINERY

sezione CC / seccion CC

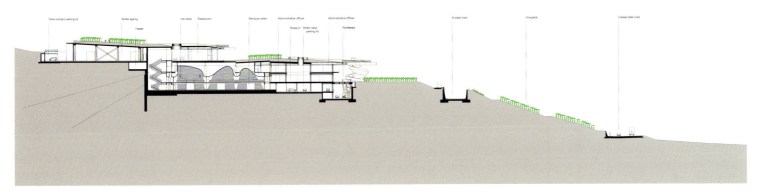

ARCHEA ASSOCIATI
CANTINA ANTINORI
ANTINORI WINERY

sezione DD / seccion DD

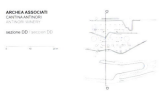
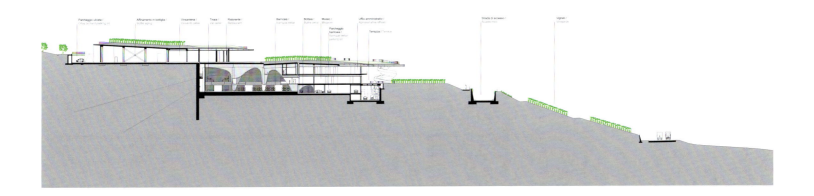

The secluded heart of the winery, where the wine matures in barrels, conveys, with its darkness and the rhythmic sequence of the terracotta vaults, the sacral dimension of a space which is hidden, not because of any desire to keep it out of sight but to guarantee the ideal thermo-hygrometric conditions for the slow maturing of the product. The architectural section reveals that the altimetric arrangement follows both the production process of the grapes and the tourist browsing route.

The materials and technologies evoke the local tradition with simplicity, coherently expressing the theme of studied naturalness, both in the use of terracotta and in the advisability of using the energy produced naturally by the earth to cool and insulate the winery, creating the ideal climatic conditions for the production of wine.

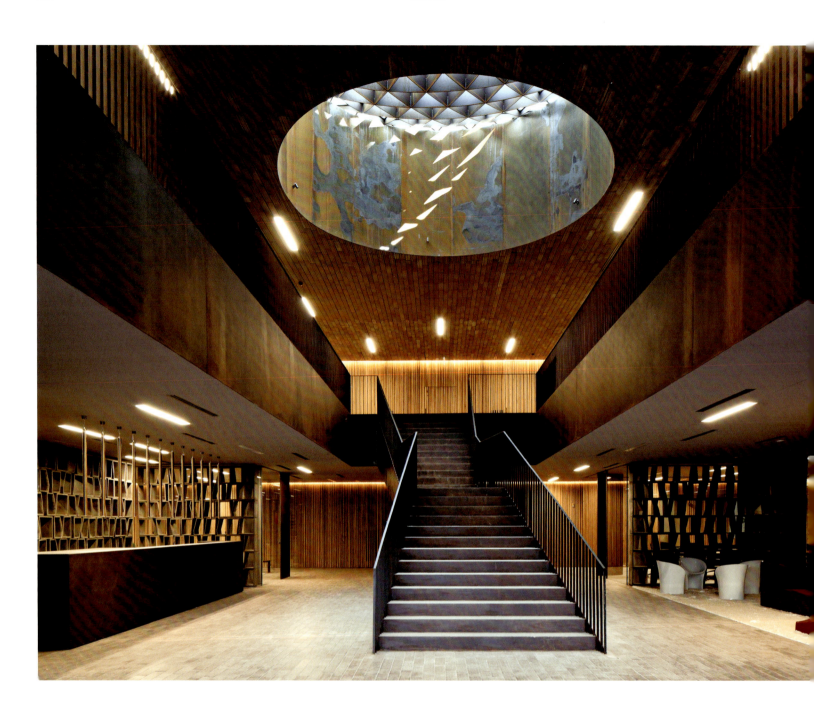

Winery Nals Margreid

Nals is located at the bottom of the "Sirmianer" hill embedded in a scenery of wine and fruit. The porphyry walls of the mountain ridge breakthrough this grape and scape and form with their dark brown-red colour a strong contrast to the lovely scenery of wine. The desire to merge the two places of production "Margreid" and "Nals" and to expand the wine production in Nals requires an advance in the capacities considering the oenological sophisticated processing of the grapes.

The new arrangement of the functions has been put into practice through the realization of a new head-end structure for the delivery and cellaring of the grapes with an adjoining vinification tower, a large new underground cellar, as well as a new barrel cellar into the court and a spanning far cantilevering roof panel. The two central parts of the winery open up towards the main court and provide the visitors with an insight into the barrel cellar and the vinification tower.

The new main building with a socket zone is built up of brown-red coloured insulating concrete, which forms a chromatic unit with the bordering cemetery consisting of porphyry stones as well as the porphyry cliff of the Sirmianer hill. The convolution on the bottom side of the shingle follows the static force lines and creates a bracing surface like an origami. The barrel cellar is placed into the court like an oversize case of wine, according to this, it is a wooden structure.

The new materials are left naturally, so they blend harmoniously into the overall setting and thus correspond with the basic ideology of the product manufactured: independent, honest and authentic.

Architects: Markus Scherer / Location: Bolzano, Italy / Client: Kellerei Nals-Margreid / Photography: Bruno Klomfar

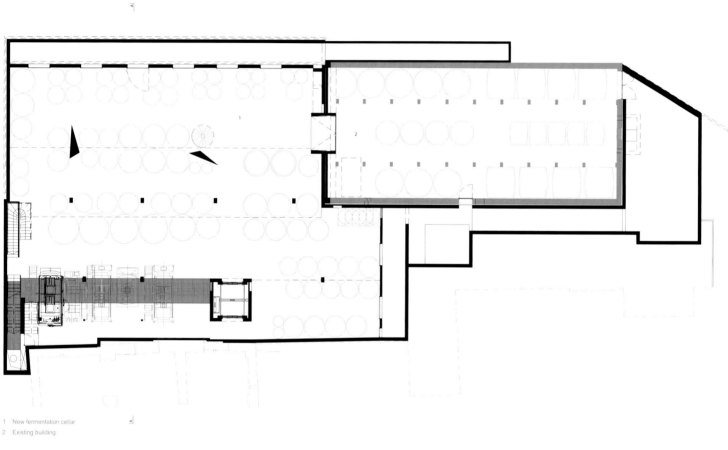

1 New fermentation cellar
2 Existing building

0 1   5   10m

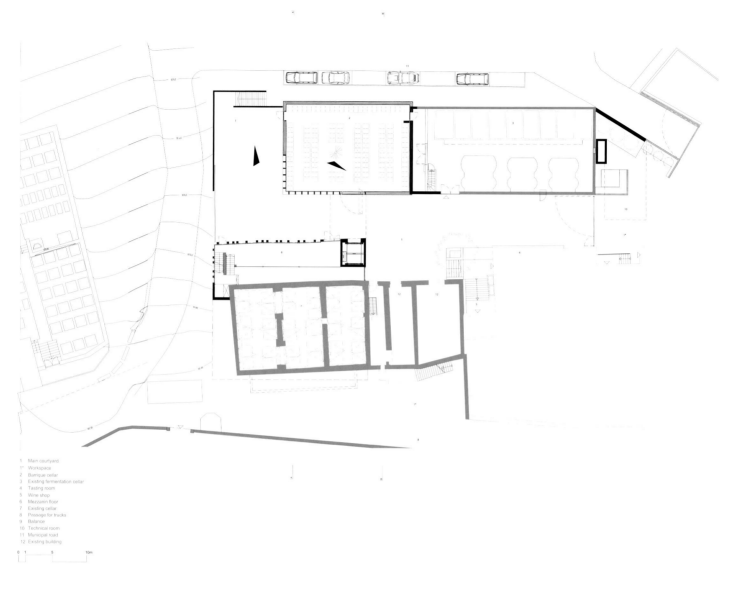

1   Main courtyard
1* Workspace
2   Barrique cellar
3   Existing fermentation cellar
4   Tasting room
5   Wine shop
6   Mezzanin floor
7   Existing cellar
8   Passage for trucks
9   Balance
10 Technical room
11 Municipal road
12 Existing building

0   1   5   10m

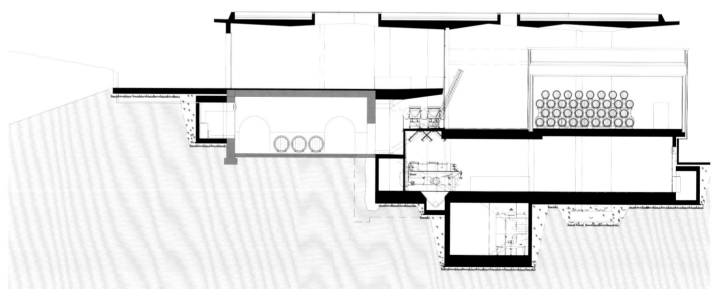

SECTION AA

0   1   5   10m

Soif d'ailleurs

"Soif d'ailleurs" is a foreign wine shop with a surface area of 170 m². It is a sales, tasting and reception area. The main constraint of the space was the low ceiling height and great depth. Architects therefore worked on the three functional sequences from the window on the street. The part on the street serves as a wine shop, the intermediate part offers tasting areas by the glass and the two rooms at the back can be used for various events and receptions.

The shop on the street is distinguished by the double height created, linking the sales area with the administrative area on the floor. The back rooms are naturally lit by a glass roof. They can be isolated from the sales area by a sliding door.

The design of the project was intended to be elegant and sober to highlight the products intended for sale. The floor is made of clear terrazzo with inlays of green glass. All the furniture in light oak was made to measure. The bar and counter tops are in white marble.

Particular care was taken in the scenographic reflection and in the scenarisation of the contents. To optimise and enrich the various bottle presentations, backlit walls allow, for example, a selection of products to be highlighted. Shelves and display cabinets in the shop area present the flagship products. Furniture elements such as wine libraries present the rest of the products on sale.

Architects: ATELA Architectes / Interior Design: ATELA Architectes / Client: Mathieu Werhung / Photography: Gitty Darugar
Materials of Wine Rack: Mainly Wood, Terrazzo for the floor, Plastic, Stainless steel, Marble

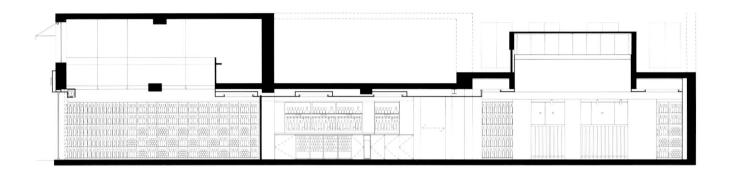

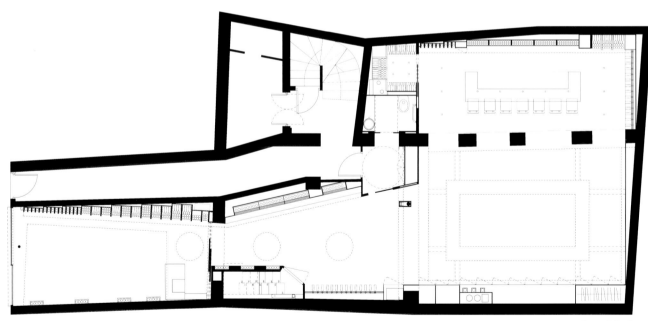

Plan RDC

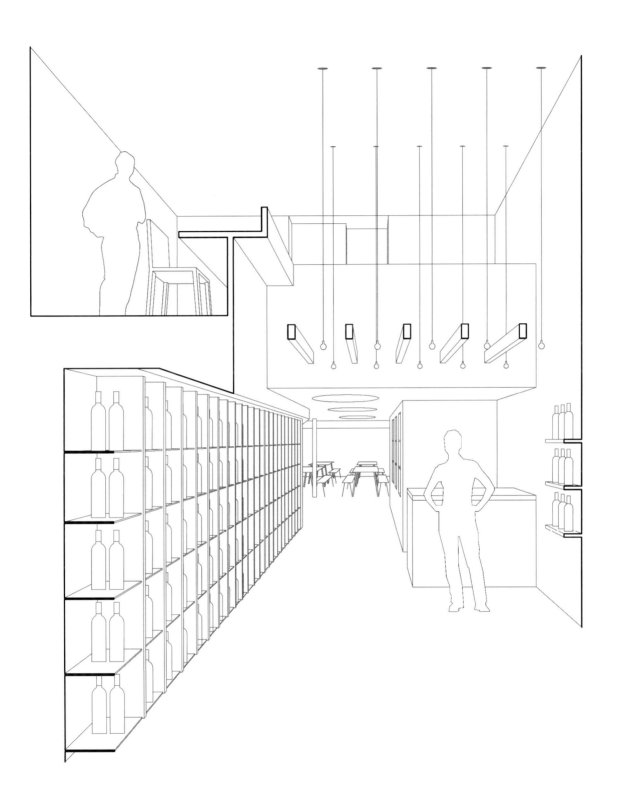

Macrolink Group Clubhouse Wine Cellar

The project is located in Building 40, South District, Macrolink Industrial Park, Tongzhou, Beijing. It is a space integrating wine, liquor, and cigars with professional cellaring and reception experience. Stepping into the wine cellar entrance, the first thing that catches your eye is the high-end collection area. A variety of excellent wine vessels are displayed on the towering and erect semi-open cylindrical showcase. A huge circular lighting lamp hangs in the centre, like a bright yellow moon in the sky, radiating a charming different light and shadow.

The walls on both sides of the high-end collection area are paved with quaint natural stone, and the high and low boutique showcases stand against the walls. The vertical illumination of the light makes it the focus of the audience, attracting visitors to find out. Bypassing the arc-shaped exhibition area, you can directly enter the liquor storage area. The solid and heavy solid wood wine rack has abandoned all superfluous decorations, and the whole box of liquor is quietly placed on it.

Behind the door on the right side of the high-end collection area is the clubhouse cigar bar. The surrounding walls are paved with the most rustic blue bricks, and five sets of boutique showcases stand along the walls. On the right side of the cigar area, the nearly 100 m^2 ultra-luxury tasting room is quietly waiting for guests to visit.

There are eight long and narrow wine storage areas symmetrically distributed on both sides like a cellar, and the rows of wine racks are quite spectacular. Curved ceilings, unique artistic lamps, elegant European-style arches, exquisite decorative paintings... It's like being in the cellar of a century-old winery.

Design Agency: Sicao Wine Cellar / Design Team: Sicao A+ design team / Project Manager: Zhonghui Jing
Location: Tongzhou, Beijing / Area: 1000 m^2 / Photography: Lightgray Studio

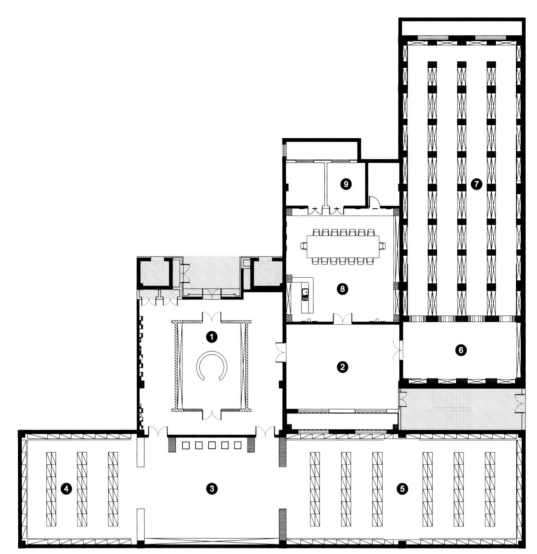

1. Corporate Culture Exhibition Area
2. Lobby
3. Excellant wine display area
4. Liquor Cellar Area A
5. Liquor Cellar Area B
6. Wine Cellar Area A
7. Wine Cellar Area B
8. Tasting area
9. warehouse

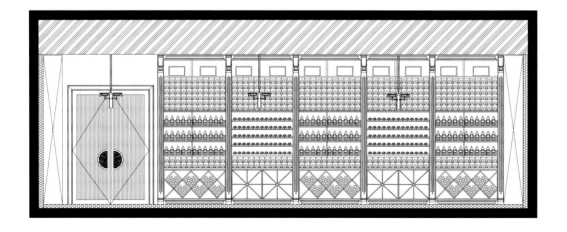

Bodegas Murviedro Showroom, Winery & Cellar

Located in the historical area of the Plaza de la Villa in Requena, the Casa de la Seda (the Silk House) is the result of a project sponsored by the City Council and Bodegas Murviedro. This project is the combination of three original houses and their respective grottos located beneath the actual plaza.

The Casa de la Seda was built to introduce its visitors to the entire process involved in wine production. With its Artisanal Winery, Museum, and Tasting Room, the House offers an integral educational experience that begins by showing how the people used to live and ends with the complete elaboration process.

In reality, the project consists of a singular Staircase, a sculpture of corten steel that spans the equivalent of four altitudes and succeeds in connecting up to eight different levels (all of which correspond to the different functions in the plan). It initiates in the Cave and reaches, in order, the Hall, the Artisanal Winery, the Store, the Tasting Room, the Exhibit, the Meeting Room and finally the Technical area.

Of noted interest is the intervention (or lack thereof) in the Caves, where a single wooden walkway of Ipe Wood wraps deviously around the natural structures and illuminates the three caves while respecting their borders and not interfering with them.

Architects: Rubén Muedra Estudio de Arquitectura / Interior Design: Rubén Muedra Estudio de Arquitectura / Location: Valencia, Spain / Area: 390 m² / Photography: Adrián Mora Maroto / Materials of Wine Rack: Wood, Metal, Corian Solid Surface

The facade.

The exhibit.

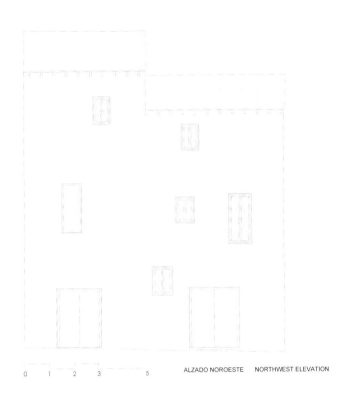

0 1 2 3 5   ALZADO NOROESTE   NORTHWEST ELEVATION

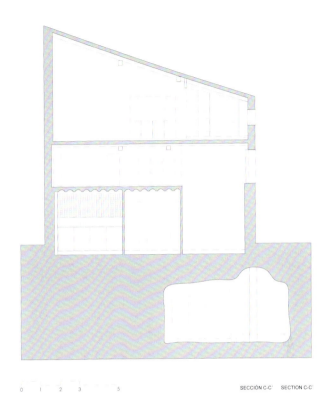

0 1 2 3 5   SECCIÓN C-C'   SECTION C-C'

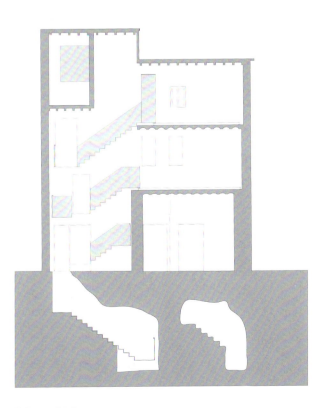

0 1 2 3 5   SECCIÓN A-A'   SECTION A-A'

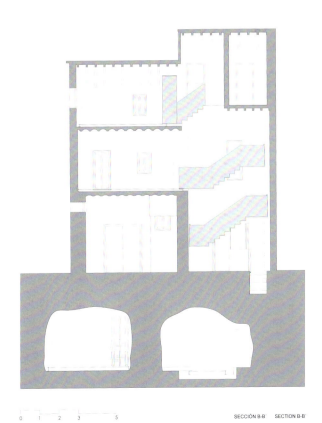

0 1 2 3 5   SECCIÓN B-B'   SECTION B-B'

A singular staircase, a sculpture of corten steel that spans the equivalent of four altitudes and succeeds in connecting up to eight different levels.

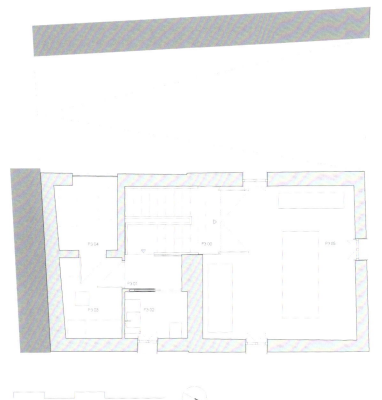

0 1 2 3   5   N

PLANTA TERCERA   THIRD FLOOR   AREA

P3 00. ESCALERA P2   P2 STAIRS   3.41 M²

P3 01. DISTRIBUIDOR   HALL   2.08 M²

P3 02. ASEO   PRIVATE TOILET   3.12 M²
        PRIVADO

P3 03. SALA EQUIPO   TECHNICAL ROOM   5.11 M²
        TÉCNICO

P3 04. SALA   FACILITIES   4.76 M²
        MAQUINARIA
        EXTERIOR

P3 05. ALMACÉN   WAREHOUSE   23.88 M²

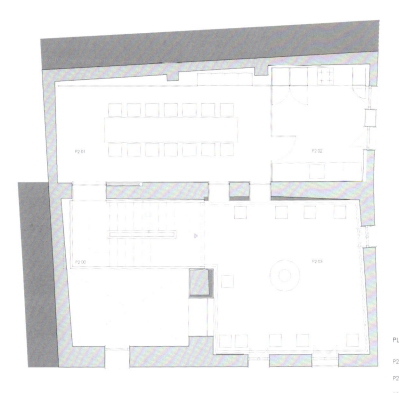

0 1 2 3   5   N

PLANTA SEGUNDA   SECOND FLOOR   AREA

P2 00. ESCALERA P1   STAIR P1   10.36 M²

P2 01. SALA DE   TASTING ROOM   22.14 M²
        CATAS

P2 02. COCINA   KITCHEN   11.45 M²

P2 03. ESPACIO   EXHIBITIONS   25.64 M²
        EXPOSITIVO   ROOM

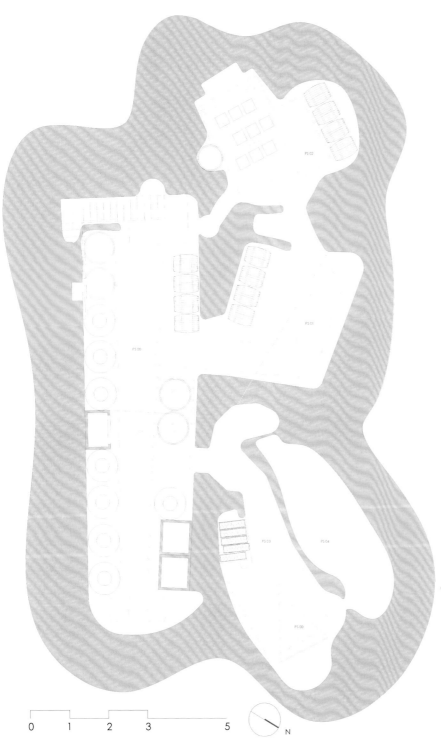

PLANTA CUEVA	CAVE FLOOR	AREA
PS 00. SALA EXPOSITIVA	**EXHIBITION ROOM**	71,94 M²
PS 01. SALA VINIFICACIÓN	VINIFICATION ROOM	27,79 M²
PS 02. ALMACÉN BARRICAS	CASK WAREHOUSE	30,64 M²
PS 03. ALMACÉN CAVA	SPARKLING WINE WAREHOUSE	27,90 M²
PS 04. SALA ANEXA	SPARKLING WINE	17,39 M²

0   1   2   3   5

N

Chateau Cheval Blanc Winery

In the Saint-Emilion vineyard and the long curves of an age-old landscape shaped along the lines of the vines by the Romans, Cheval Blanc is a small property, along which the cellar, where one of the best wines in the world is made, was a poorly functioning casemate. In 2011, the owners Bernard Arnault and Baron Albert Frère asked Christian de Portzamparc to build a new winery. The new cellar is installed in the extension of the volume and colour of the stones of the castle in a movement of curved sails that starts from the ground to become a hill, a promontory, a belvedere to see far away. The four long concrete sails act as walls and beams and house the large vat room. A large passage through the vineyard houses the reception of the harvest.

For the vat room, when others turn to wood or stainless steel, Pierre Lurton has chosen to keep the principle of concrete for vinification because of its thermal inertia and the reaction of the walls to tartaric acids. The concrete tanks have been designed in a design of jars that widen in their core to optimize the oxygenation surface. The tanks are at the heart of the architecture, and respond to the concrete sails. Zenithal lighting lines provide the interior with a softness of natural light that descends and sculpts the curved surfaces and tanks. Fifty-two vats, in nine different sizes, make it possible to vinify very precisely according to the size of each plot and the height of the harvest.

In the basement, the livestock cellar is like a crypt, with another light lined with openwork brick walls. She sees dozens of barrels lining up between the inside and the outside, the cellar is a place of mutation and interaction with nature, the air and the climate during the seasons. No line here is superfluous, everything is involved in the wine making process and its gestures. Here, architecture joins the excellence of a very old experience to improve the production of exceptional wine.

Architects: Christian De Portzamparc / Area: 5,250 m^2 / Client: Château Cheval Blanc

View of the vineyard of Saint-Emilion

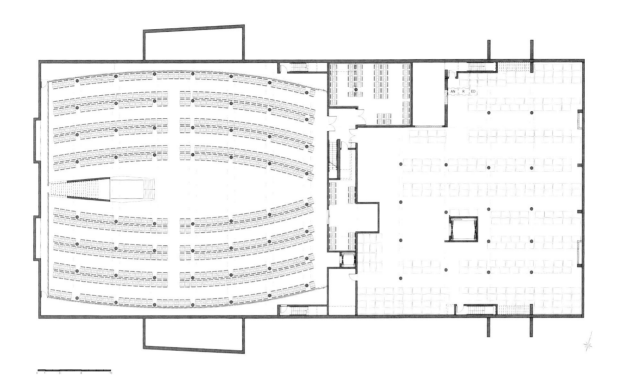

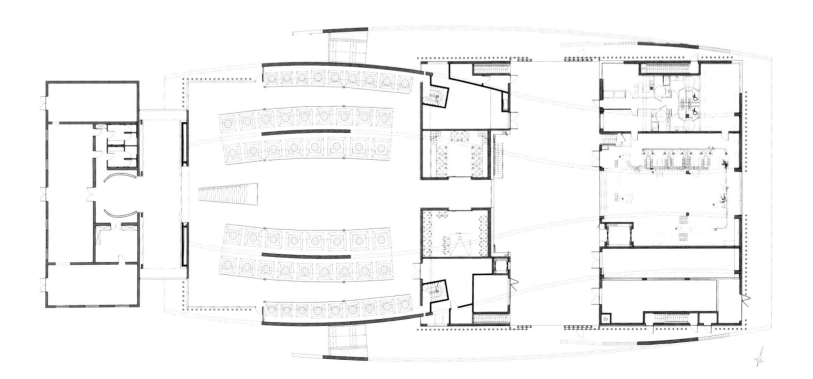

The 52 concrete tanks have been designed in a design of jars that widen in their core to optimize the oxygenation surface.

Zenithal lighting lines provide the interior with a softness of natural light that descends and sculpts the curved surfaces and tanks.

In the basement, the livestock cellar is like a crypt, with another light lined with openwork brick walls.

Huawei Wine Cellar

This design is a constant temperature wine cellar for Suzhou Huawei by Shenzhen Raching Technology. Its design is a fashionable, simple yet spectacular European style. This wine cellar has an arched ceiling, installed with a traditional European ceiling lamp. The softness of lamp lights perfectly matches the lustre of this wine cellar. It silently creates a luxurious atmosphere for wines and reflects the elegance of wines.

All the 6 walls inside this wine cellar were treated with temperature insulation and moisture proof process. Surfaces of the wine cellar walls and ceiling are lacquered with moisture-proof diatom coating material. The wine cellar has a parquet floor made of marble material. The wine display shelves are seamlessly connected to the wine cellar walls. The shelves are designed with storage units in different directions, obliquely, horizontally, vertically, to ensure the shelves could store various shapes of wines bottles, including champagne, 1.5L/3L wines etc.

Design Agency: Shenzhen Raching Technology / Design Team: Shenzhen Raching Technology / Location: Jiangsu, China

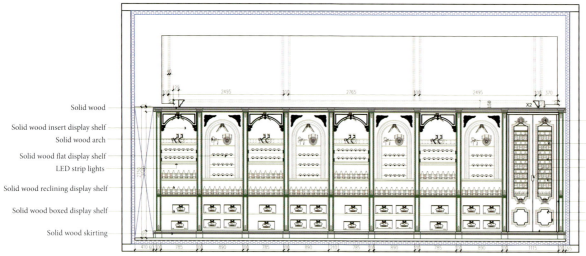

Solid wood

Solid wood insert display shelf

Solid wood arch

Solid wood flat display shelf

LED strip lights

Solid wood reclining display shelf

Solid wood boxed display shelf

Solid wood skirting

Solid wood carving

Wooden thermostatic champagne cooler

Hollow electrically heated glass

Concealed hand pull

Solid wood reclining display shelf

Wooden cabinet with insulated sides

Wooden door trim

Solid wood lines

Door lock

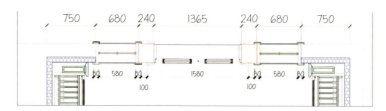

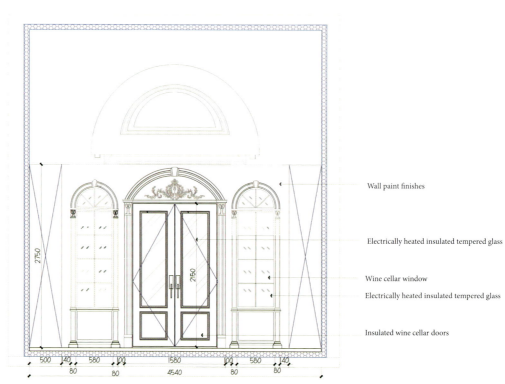

Wall paint finishes

Electrically heated insulated tempered glass

Wine cellar window

Electrically heated insulated tempered glass

Insulated wine cellar doors

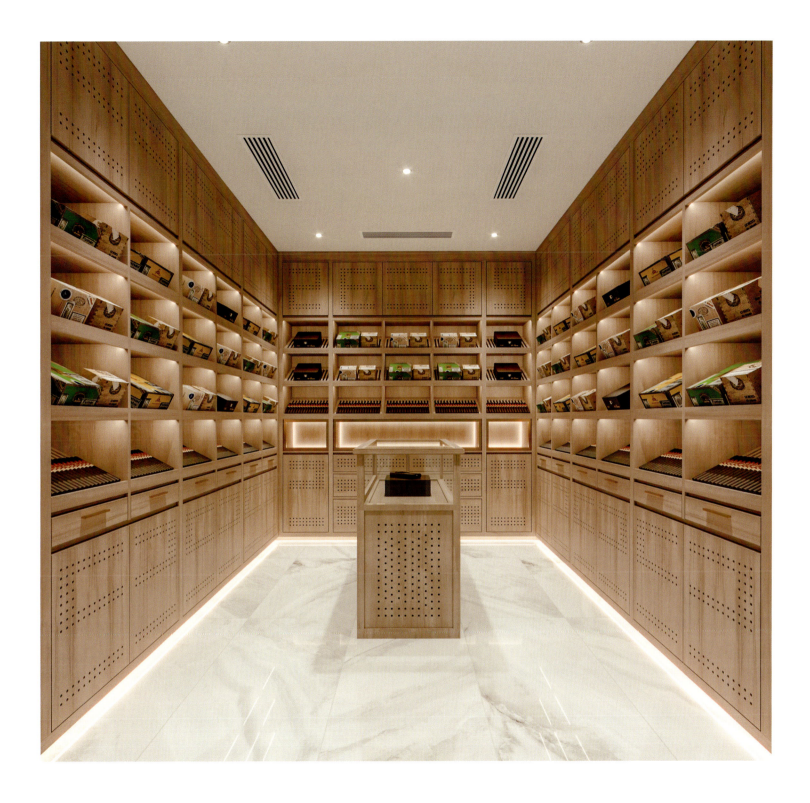

Offering perfect constant temperature and steady 70% humidity while adopting entire Spanish cedar wood, Raching customized cigar Humidor Rooms could let cigars be soft and smooth, just as those cigars finish SPA and stay at a cozy cottage.

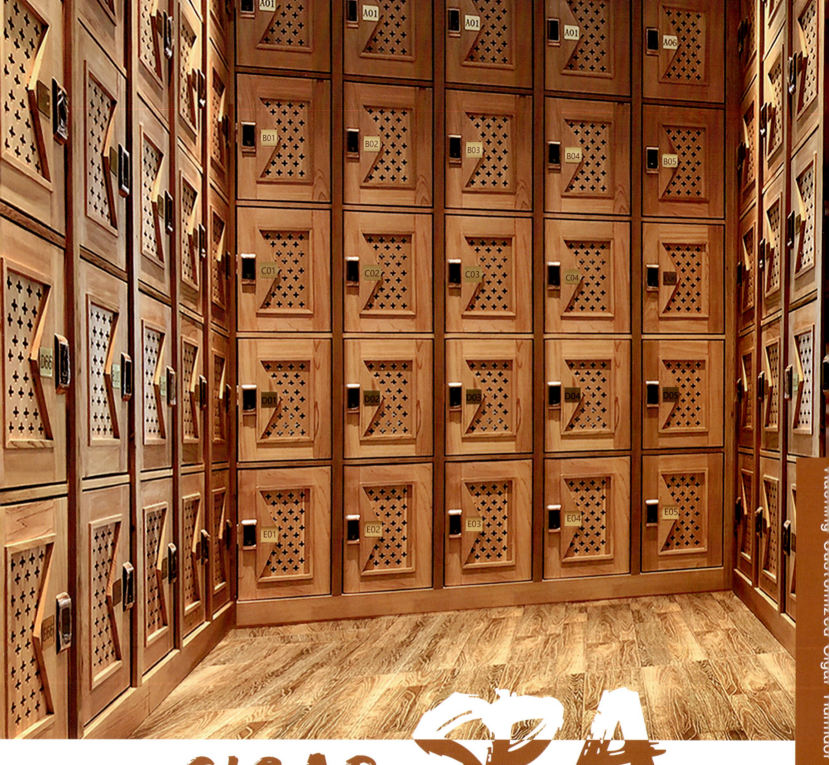

CONTRIBUTION

ARTPOWER

Acknowledgements

We would like to thank all the designers and companies who made significant contributions to the compilation of this book. Without them, this project would not have been possible. We would also like to thank many others whose names did not appear on the credits, but made specific input and support for the project from beginning to end.

Future Editions

If you would like to contribute to the next edition of Artpower, please email us your details to: press@artpower.com.cn